Upgrading India's Electronics Manufacturing Industry: Regulatory Reform and Industrial Policy

DIETER ERNST

EAST-WEST CENTER
COLLABORATION · EXPERTISE · LEADERSHIP

EAST-WEST CENTER
COLLABORATION • EXPERTISE • LEADERSHIP

The East-West Center promotes better relations and understanding among the people and nations of the United States, Asia, and the Pacific through cooperative study, research, and dialogue. Established by the US Congress in 1960, the Center serves as a resource for information and analysis on critical issues of common concern, bringing people together to exchange views, build expertise, and develop policy options.

The Center's 21-acre Honolulu campus, adjacent to the University of Hawai'i at Mānoa, is located midway between Asia and the US mainland and features research, residential, and international conference facilities. The Center's Washington, DC, office focuses on preparing the United States for an era of growing Asia Pacific prominence.

EastWestCenter.org

For information or to order copies, please contact:

Publication Sales Office
East-West Center
1601 East-West Road
Honolulu, Hawai'i 96848-1601

Tel: 808.944.7145
Fax: 808.944.7376

EWCBooks@EastWestCenter.org
EastWestCenter.org/Publications

ISBN: 978-0-86638-244-1 (print) and 978-0-86638-245-8 (electronic)

© 2014 East-West Center

Upgrading India's Electronics Manufacturing Industry: Regulatory Reform and Industrial Policy
Dieter Ernst

Abstract

India faces a fundamental puzzle. The country is a leading exporter of information-technology services, including knowledge-intensive chip design. Yet electronics manufacturing in India is struggling despite a huge and growing domestic market and pockets of world-class capabilities.

To examine this puzzle the World Bank commissioned this study in May 2013 on behalf of the Chief Economic Advisor, Government of India, Raghuram Rajan (now the governor of the Reserve Bank of India). Drawing on extensive survey questionnaires and interviews with key industry players (both domestic and foreign) and relevant government agencies, this study identifies major challenges India-based companies face in engaging in electronics manufacturing. The analysis culminates in detailed policy suggestions for regulatory reform and support policies needed to unblock barriers to investment in this industry and to fast-track its upgrading through innovation.

This study finds that restrictive regulations and a largely dysfunctional implementation of past support policies have constrained investment in plants and equipment and technology absorption and innovation. India's strength in chip design does not help. Local electronics manufacturing remains disconnected from India's chip-design capabilities which are integrated, instead, into global networks of innovation and production. India's growing domestic demand for electronic products results in rising imports of final products and high import-dependence for key components. These imports have become the third-most-important driver, after petroleum and gold, of the country's record current-account deficit.

Bold action is required to change the anemic growth of electronics manufacturing as well as change outdated patterns of policy responses. This is particularly true in light of the fast-changing dynamics in the electronics industry and new challenges resulting from international trade agreements such as the Information Technology Agreement. Unlike China and earlier industrial latecomers from Asia, India can no longer rely exclusively on "high-volume, low-cost" manufacturing for rapid growth in electronics manufacturing. India also needs to pursue a niche-market strategy focusing on "low-volume, high-value" products.

Accumulated strengths in electronic systems and integrated-circuit design could provide the basis for such "low-volume, high-value" electronics manufacturing. But for this to happen, India would need to link circuit design-and-development capabilities, now trapped within the research-and-development laboratories of multinational-corporation affiliates, back to India-based companies serving India's domestic markets. That task is made more difficult by India's fragmented innovation system characterized by weak links between education, research, and industry—a challenge of which many in India are acutely aware.

Domestic challenges constrain the capacity for the productivity-enhancing innovation of India's electronics industry just when the global electronics industry is rapidly ending historical strategies for growth.

To achieve its potential, electronics manufacturing in India needs an adjustment in its industrial-growth model. The industry must move from *fragmentation*, chasing "high-volume, low-cost" activities, towards *integration*, with a greater focus on "low-volume, high-value" and, for high volume, on frugal innovation for the domestic market.

The government's National Policy on Electronics is an important first step on this path. The plan must be communicated directly to manufacturing companies and implementation needs to be focused on its key components and sustained over the coming years. The plan then needs to be complemented, on the one hand, by reforms beyond the industry, especially those relating to taxation, customs, compliance, and inspections; and, on the other, by process changes, especially the strategic use of technical standards.

Policies to upgrade India's electronics manufacturing industry need to place considerable effort on developing smart approaches to international trade diplomacy that go beyond tariff reductions and address the increasing importance of technical barriers to trade. Technical barriers to trade include standards and the unequal distribution of financial gains generated by trade among countries differing substantially in their stage of development and in their economic institutions and capabilities.

India's engagement with the institutions shaping trade and investment in this industry is a critically important complement to the more high-profile efforts to build domestic production (including semiconductor wafer fabrication facilities, or "fabs," for leading-edge semiconductors). Such a two-pronged strategy would likely provide an enduring and sustainable boost to the electronics manufacturing industry in India.

Table of Contents

List of Acronyms

3DP	(*also* 3D printing) three-dimensional printing
ABB	ABB Group
ACTA	Anti-Counterfeiting Trade Agreement
AMD	Advanced Micro Devices
ASEAN	Association of Southeast Asian Nations
ASEAN-5	Indonesia, Malaysia, the Philippines, Singapore, and Thailand
ASSOCHAM	Associated Chambers of Commerce and Industry of India
BEL	Bharat Electronics Limited
BIS	Bureau of Indian Standards
BSNL	Bharat Sanchar Nigam Limited
CAD	current-account deficit
CASCO	Committee on conformity assessment at ISO
CASPA	Chinese American Semiconductor Professional Association
CEAMA	Consumer Electronics and Appliances Manufacturers Association
CIER	Chunghua Institution of Economic Research
CII	Confederation of Indian Industry
CR	concentration ratio
CRT	cathode ray tube
CST	Central Sales Tax
CVD	countervailing duty
DEITy	Department of Electronics and Information Technology
DGFT	Directorate General of Foreign Trade
DIN	(Germany's) Institute for Standardization
DOE	Department of Electronics
DOSTI	Development Organization of Standards for Telecommunications in India
DST	Department of Science and Technology
ECG	electrocardiogram
ECIL	Electronic Corporation of India Ltd
EDA	electronic design automation
EDF	Electronics Development Fund
EITDC	Electronics and Information Technology Division Council
ELCINA	Electronic Industries Association of India
EMC	electronics manufacturing cluster
EMS	electronic manufacturing service
ERSO	Electronics Research and Service Organization
ESD	electrostatic discharge
ESDM	Electronic System Design and Manufacturing
ETSI	European Telecommunications Standards Institute
FDI	foreign direct investment

FPD	flat-panel displays
FTA	free trade agreement
GE	General Electric
GIN	global innovation network
GISFI	Global ICT Standardization Forum for India
GPA	Government Procurement Agreement
GST	Goods and Services Tax
HP	Hewlett-Packard
IBIDEN	IBIDEN Co., Ltd. (*formerly* Ibigawa Electric Industry Co., Ltd.)
IBM	International Business Machines Corporation
IC	integrated circuit
ICT	information and communications technology
IEC	International Electrotechnical Commission
IEEE-SA	Institute of Electrical and Electronics Engineers Standards Association
IESA	Indian Electronics and Semiconductor Association
IIM	Indian Institute of Management
IISC	Indian Institute of Science
IIT	Indian Institute of Technology
IMF	International Monetary Fund
INSEAD	International Business School in Fontainebleau, France
IP	intellectual property
IPR	intellectual property rights
IS	Indian Standards
ISO	International Organization for Standardization
IT	information technology
ITA	Information Technology Agreement
ITRI	Industrial Technology Research Institute
LCD	liquid-crystal display
LED	light emitting diode
LG	LG Electronics (*formerly* GoldStar)
MAI	Market Access Initiative
MAIT	Manufacturers' Association for Information Technology
MDA	Market Development Assistance
MCO	multi-component integrated circuit
MFN	most favored nation
MNC	multinational corporation
MOEA	(Taiwan's) Ministry of Economic Affairs
M-SIPS	Modified Special Incentive Package Scheme
MSMEs	micro, small, and medium enterprises
NABCB	National Accreditation Board for Certification Bodies
NABL	National Accreditation Board for Testing and Calibration Laboratories
NPE	National Policy on Electronics
NQC	National Quality Control
NTB	non-tariff barrier

NTP	National Telecom Policy
OECD	Organisation for Economic Co-operation and Development
OEM	original equipment manufacturer
PCB	printed circuit board
PMA	preferential market access
PSU	public sector units
PTA	preferential trade arrangements
PV	photovoltaic
Q	"quarter," as in a quarter of a calendar or fiscal year
QCI	Quality Council of India
RIM	Research In Motion Limited (*now* BlackBerry Limited)
RoHS	Restriction of Hazardous Substances
SCL	Semiconductor Complex Limited
SEZ	special economic zone
SDO	standards-development organization
SHIS	Status Holder Incentive Scrip
SITP	Scheme for Integrated Textile Parks
SMEs	small and medium enterprises
SOE	state-owned enterprise
SMT	surface-mounted technologies
STB	set-top box
TBT	technical barriers to trade
TCL	TCL Communication Technology Holdings Limited
TEC	Telecommunications Engineering Center
TIER	Taiwan Institute of Economic Research
TPP	Trans-Pacific Partnership Agreement
TRIMS	Agreement on Trade-Related Investment Measures
TRIPS	Trade Related Aspects of Intellectual Property Rights
TSIA	Taiwan Semiconductor Industry Association
TTIA	Transatlantic Trade and Investment Agreement
USITC	United States International Trade Commission
WIPO	World Intellectual Property Organization
WTO	World Trade Organization
ZTE	ZTE Corporation (*formerly* Zhongxing Semiconductor Co. Ltd.)

Executive Summary

CONTEXT

Surging Demand, Struggling Supply

Electronics manufacturing in India is struggling despite a huge and still-growing domestic market and pockets of world-class capabilities in information technology services and chip design. Local production faces cost disadvantages which constrain investment in plants and equipment, technology absorption, and innovation. Local production hardly benefits from India's chip-design capabilities which are integrated, instead, into global multinational corporation (MNC) networks of innovation and production.

India's growing domestic demand for electronic products generates rising imports of final products and high import-dependence for key manufacturing components. These imports have become the third-most-important driver, after petroleum and gold, of the country's record current-account deficit.

Unless these fundamental weaknesses are addressed soon, the electronics manufacturing sector is unlikely to achieve the targets set in the nation's Twelfth Five-Year Plan: output of US$120 bn and millions of jobs by 2017. Bold action is required to initiate a break with the anemic growth of electronics manufacturing as well as with outdated patterns of policy responses. This is particularly true in light of the fast-changing dynamics in the electronics industry.

Vertical Specialization Defines India's New Manufacturing Imperative

Unlike China, India can no longer rely exclusively on "high-volume, low-cost" electronics manufacturing, India must also pursue a niche-market strategy focusing on "low-volume, high-value" products

Such action must first take into account the manufacturing imperatives created by a fast-changing industry with more change on the horizon. Geographically dispersed networks of production and innovation have fragmented electronics manufacturing. Product-life cycles have been and are being drastically reduced. China's "high-volume, low-cost" manufacturing locations have created capabilities to rapidly scale up new production lines—but even they are struggling to keep up with the pace of technological change. At the same time, advanced manufacturing technologies, especially additive manufacturing (often called "3D printing") may facilitate mass customization based on "low-volume, high-value" production. This may challenge existing distributions of competitive advantage.

These trends create an important strategic challenge for India's electronics industry. Unlike China and earlier industrial latecomers from Asia, India can no longer rely exclusively on "high-volume, low-cost" manufacturing for rapid growth in electronics manufacturing. India must also pursue a niche-market strategy focusing on "low-volume, high-value" products.

India would seem to be well qualified to address this imperative. Accumulated strengths in electronic systems and integrated-circuit design could provide the basis for such "low-volume, high-value" electronics manufacturing. However, deep integration of Indian electronic-design capabilities with global research-and-development (R&D) networks has produced little integration of these capabilities with India's own domestic electronics manufacturing value chain.

To reap the benefits of value-chain integration, India needs to link circuit-design-and-development capabilities, now trapped within the R&D laboratories of MNC affiliates, back to India-based companies serving India's domestic markets. This is made difficult by India's fragmented innovation system—characterized by weak links between education, research, and industry—a challenge of which many in India are acutely aware.

India thus faces a fundamental challenge: Domestic challenges constrain the capacity for the productivity-enhancing innovation of India's electronics industry just when the global electronics industry is rapidly ending historical strategies for growth. The electronics industry has recognized these constraints and policymakers have begun to respond. To evaluate those responses, however, it is important to understand the constraints imposed by the global market structure, particularly those defined by trade policy and global oligopolies.

POLICY PARAMETERS

Trade Policy

India's experience with trade liberalization through international trade agreements has had two sides. Some sectors—such as information technology (IT) services, car components, and generic pharmaceuticals—have benefitted from India's membership in the World Trade Organization (WTO). As far as electronics manufacturing is concerned, however, WTO membership obliges India to ensure "compliance" of its industrial and innovation policies with increasingly complex trade rules. Current rules constrain India's options for the type of national-support policies earlier available to Japan, Korea, and Taiwan.

The most important of these rules are defined by the Information Technology Agreement (ITA). This study shows that, in India's experience with the ITA, the gains from trade liberalization have been overshadowed by substantial costs—especially stalled or declining domestic electronics production.

In 1997 India joined the ITA from a position of weakness—with its electronics sector liberalized barely a year earlier and still finding its feet. With an inverted tariff structure in place thereafter, finished products then being duty free but their components not, domestic production had little chance of building capabilities or investing at sufficient scale.

In 2003 China, by contrast, joined the ITA from a position of strength. When China entered the ITA, six years after India did so, China was already the third-largest exporter and the fourth-largest importer of ITA products.

As a result of India's entry into the ITA, India's imports of key electronics products have grown much faster than domestic production. Imports now account for almost two-thirds of India's consumption of electronics products. India's imports of integrated circuits and other core electronics components grew especially quickly. The value-added portion of Indian electronics manufacturing is now less than 10 percent.

xii | EAST-WEST CENTER

The early exposure of an electronics sector unready to face the full competitive pressures of a globalized industry, already facing the effects of the inverted tariff structure, has been further exacerbated by the erection of non-tariff barriers in developed markets. These non-tariff barriers may neutralize any positive effects of ITA-induced tariff reductions in target markets.

These trade constraints are amplified by the structure of the global electronics market. That global market, far from being a field of unfettered competition, has, over the last twenty years, become more and more concentrated in oligopolies.

Global Oligopolies

A market segment is said to be controlled by a loose oligopoly when the four largest companies in the segment achieve more than 25 percent of total segment sales and controlled by a firm oligopoly when this ratio rises above 50 percent. Over the last two decades one after another segment of the global electronics industry—from personal computers (PCs) to hard disks to smartphones—has emerged as a loose or firm global oligopoly.

A handful of MNCs dominate India's electronics markets as oligopolists without engaging in substantial domestic manufacturing in India (either directly or through contractors), with the exception of low-value-added final assembly.

Oligopolies do not necessarily mean an absence of competition—often the opposite (such as the current competition between Apple and Google/Samsung). But it is important to understand the barriers to entry for electronics manufacturing in India these oligopolies create, whether such manufacturing would be by themselves or by Indian challengers.

These MNC oligopolies can rely on their extended global production networks to source the relevant products for the Indian market from their preferred, primarily Chinese, production sites. In addition to cost advantages, what matters most for MNCs is that they can benefit from the accumulated capabilities in China for rapid and low-cost up scaling up of sophisticated production lines. These capabilities to scale-up at speed imply Indian companies would not only need to match Chinese prices but to beat them, perhaps by as much as 15 percent.

Global oligopolists can erect high entry barriers for Indian companies who might seek to enter or re-enter the electronics industry. Global oligopolists can set lower prices than challengers—not only because they can source the relevant products from low-cost production sites through their global production and innovation networks but also because of their control over leading-edge technology and their superior innovation capacity.

Oligopolistic control gives rise to a "commoditization" of electronics products across the globe, imposing substantial constraints on local innovation efforts that would seek to address specific needs of India's domestic market through "frugal innovation." Successful entry into those markets would require quite extraordinary efforts by Indian companies to develop superior business models and new technologies. For that to happen the Indian government and the Indian private sector would need to join forces and develop a decisively longer-term industrial-development strategy combining smart regulatory reform and structural support for electronics industries.

A handful of MNCs dominate India's electronics markets as oligopolists without engaging in substantial domestic manufacturing in India

RESPONSES & RECOMMENDATIONS

These findings are elaborated in detail in the main body of this study and should not be taken to mean that international factors dominate domestic ones. The constraints on electronics manufacturing in India are as much made within the nation as made abroad. India's policymakers and the industry's stakeholders have many pathways to overcoming them—at least enough to catalyze growth. Regulatory reform together with a sustained set of industrial support policies could help to quickly unblock many barriers to investment and growth in India's electronics manufacturing.

Many of the elements of such strategies are known to domestic companies and to policymakers. The government of India has already incorporated many such elements in the Twelfth Five-Year Plan as well as in the National Policy on Electronics (NPE) now in the initial stages of implementation.

However, this study's survey of electronics companies shows that many remain either unfamiliar with the NPE or skeptical of the details of its implementation. On the other hand, a range of industry associations were involved in its formulation and their leaders did know it intimately. This may imply that communication within some of the associations needs strengthening but also indicates that the next stage in improving industrial dialogue is to reach down to individual companies.

In that vein, this study complements the NPE on two fronts: a) *specific policy recommendations* reaching beyond electronics manufacturing to the business environment in general, including tax policies and regulations, and b) *fundamental process changes*—such as in industrial dialogue, standards, and trade diplomacy—to improve policy outcomes over the long term.

These implications are summarized under the following three headings. While discrete changes in regulation or support policies (I) should be a starting point, parallel efforts under headings II and III are required to enhance the impact and sustainability of such policies.

I.   Discrete Changes in Regulation or Support Policies

1. Necessary first steps: quickly unblock barriers to investment through regulatory simplification and national market integration. In particular:

   - Speedy transition to a unified Goods and Services Tax (GST) system, the single most commonly cited "reform wish" from electronics manufacturers

   - Drastic simplifications in the business regulatory environment, in particular on dispute resolution for customs conflicts as well as formal and informal penalties for growth

2. Devise and enforce quality standards on high-priority products (e.g., medical devices, set-top boxes) to protect against dumping

3. Reduce or remove tariffs all the way up the supply chain to remove the inverted tariff structure. This requires addressing domestic and international implementation constraints:

   - Conflicts of interest between one segment of the value chain and another and between the central government and state governments

Regulatory reform together with sustained industrial support policies could help to quickly unblock many barriers to investment and growth in India's electronics manufacturing

- Constraints from existing international trade agreements (especially the Information Technology Agreement)

4. Reduce short-term infrastructure bottlenecks such as power and transportation and foster specialized electronics manufacturing clusters consistent with the demands from companies for appropriate locations.

5. Concerted effort to strengthen both vocational training and academic curricula to achieve the higher-level skills required for electronics manufacturing.

II. Facilitate Policy Implementation Through Process Changes and Institutional Innovations

1. Focus, simplify, and improve communication through transparent and user-friendly support policies (e.g., initiating communications concerning the NPE directly with manufacturing companies)

2. Encourage "industrial dialogues" involving not only large flagship companies, but also:
   - Young companies seeking to create and commercialize new products and processes
   - University and public R&D laboratories
   - Industry associations seeking to enhance the scope for such dialogues

3. Link these participants not only to "talk shops" but to meaningful, action-oriented, committees (e.g., review the composition of the NPE assessment committees)

4. Strengthen India's capacity to develop critical technical standards (especially for inter-operability) and to improve the development and management of standards-essential patents.

III. Overhaul International Investment and Trade Diplomacy

1. Shift the environment for foreign direct investment (FDI) from zero-sum to positive-sum by combining:
   - Reduction of de facto barriers to FDI, such as fiscal and policy uncertainty
   - Incentives for foreign companies to engage in industrial upgrading, over and above mere final assembly
   - Incentives to integrate India-based electronic-design capabilities with domestic electronics manufacturing
   - Monitoring and problem-solving processes and institutions to ensure and facilitate such upgrading and linking (e.g., through the restructured industrial dialogues mentioned above)

2. Use India's strong position in the WTO to co-shape the design of a "New ITA" (and other plurilateral trade agreements and free trade agreements) beyond the entrenched defensive positions of the major trading powers
   - Improve the distribution of benefits from international trade agreements through "special and differentiated treatment" requirements

Encourage "industrial dialogues" involving especially young companies which seek to create and commercialize new products and processes

- Request a reform of ITA reflecting the reality that ITA participants differ in their stages of development, their institutions, and their resources and capabilities

- Extend ITA and other plurilateral trade agreements to include non-tariff barriers (NTBs) and technical barriers to trade (TBT)

3. Strengthen India's participation in international standards-development organizations and international standards for electronics manufacturing.

CONCLUSION

India's electronics manufacturing industry still has enormous potential as current global trends, now making earlier strategies unavailable, may open up alternate strategies benefitting India's strengths—notably its strong base of high-end capabilities and its large domestic market. To achieve its potential, though, electronics manufacturing in India needs to adjust its industrial-growth model from one of *fragmentation*—chasing "high-volume, low-cost" activities—to one of *integration*—a greater focus on "low-volume, high-value" production.

The NPE is an important first step on this path. It needs to be communicated directly to manufacturing companies and its implementation needs to be focused on its key components and sustained over the coming years. It then needs to be complemented, on the one hand, by reforms beyond the industry, especially those relating to GST, customs, and compliance and inspections; and, on the other hand, by process changes, especially the strategic use of standards.

Consider medical equipment. India has demonstrated the capacity for "frugal innovation" in this field—perhaps most famously with the development of General Electric's low-cost electrocardiogram (ECG) which subsequently disrupted developed markets. India has a wide base of medical-equipment resources—from its life-sciences researchers to its pharmaceutical companies.

In one scenario this medical-equipment eco-system could remain fragmented, offered some incentives but swamped by low-quality dumping. In another, these parts could be integrated into a thriving new industry if India succeeds in adopting robust quality standards for low-cost as well as for high-end medical devices. In the second scenario, young start-up businesses could be released from constraints on growth and enjoy foreign MNCs investing in domestic Indian R&D and manufacturing.

Currently there are signs showing conflicting aspects of both possibilities. As of this study, the NPE is pushing the development of twelve standards for medical equipment while, at the same time, requiring this to still be done through an institutional structure mandating discussion between two union cabinet ministers to define the font size on the labels of such equipment.[1]

Overall, the NPE—and the sophisticated process that produced it—is a first step in the direction of the more hopeful scenario. But the NPE urgently needs buttressing with broader reforms and permanent process improvements which will expand into other domains (particularly trade diplomacy and standards). With the NPE's sustained implementation, however, complemented by broader reforms and process changes, an upgrading of electronics manufacturing in India *is* possible.

The NPE urgently needs buttressing with broader reforms (particularly trade diplomacy and standards)

Upgrading India's Electronics Manufacturing Industry: Regulatory Reform and Industrial Policy

Electronics Manufacturing in India Lags Well Behind its Potential

A STARTING POINT OF WEAKNESS

Compared to its main competitors, India's electronics manufacturing industry is struggling. Local production faces substantial cost disadvantages ("disabilities") constraining investment in plants and equipment, technology absorption, capability development, and innovation. There is a huge gap between the rapid growth of domestic demand and the nearly stagnant domestic production—and this gap is projected to increase further (see Figure 1 below).[2]

There is a huge gap between the rapid growth of domestic demand and the nearly stagnant domestic production

Figure 1. Projected Demand-Supply Gap in Electronics Industry (USD bn)

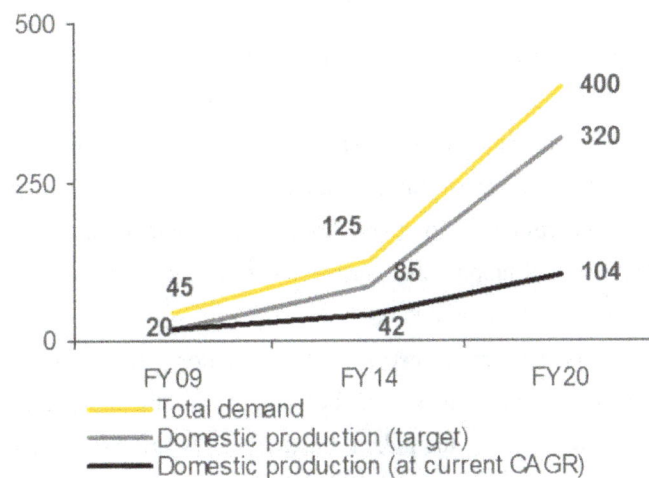

Note: "CAGR" is the "Compound Annual Growth Rate"

At the current growth rate, the demand-supply gap is projected to increase from US$25 bn in fiscal year (FY) 2009 to US$298 bn in FY2020. Such a growing gap is unsustainable—the result would be an increase in India's trade deficit to US$323 bn by 2020, equaling 16 percent of GDP.[3] To reach a value of US$400 bn in FY2020, domestic production would need to grow by 31 percent annually for the FY2009–2020 period.

Given the weakness of domestic production, India's growing domestic demand for electronic products results in rising imports. Bottlenecks abound throughout the Indian electronics industry's value chain, creating excessive import-dependence for key components.

Imports accounted for 63.6 percent of India's consumption of electronics products and 51 percent for electronic components in 2011. By 2015 these shares of imports are expected to increase to 65 percent and 61 percent respectively.[4] According to the latest Indian Ministry of Commerce and Industry data, electronic imports have increased by almost 21 percent from FY2011 to FY2012.[5]

The electronics industry has become the third-most-important driver, after petroleum and gold, of the country's record current-account deficit (see Table 1 below).

The electronics industry has become the third-most-important driver, after petroleum and gold, of the country's record current-account deficit

Table 1. India's Imports, 12 Months to March 2013 (tn rupees)

Import type	Trillion rupees
Petroleum, crude & products	9.2
Gold	2.9
Electronic goods	1.7
Machinery	1.5
Pearls, precious & semiprecious stones	1.2

Source: Directorate General of Foreign Trade (GFT) India data, as quoted in Thomson Reuters Datastream

A Narrow and Eroding Domestic Component Base

A patchy value chain limits the scope for expanding and upgrading India's electronics manufacturing industry. While India has significant capabilities in digital integrated-circuit (IC) design, most of these capabilities are not linked to the domestic market.[6] India lacks strong capabilities in semiconductor fabrication, component manufacturing, system design, and systems manufacturing and supply-chain management. Each of these weaknesses will be examined in some detail later in this study.

A particular concern is a narrow and eroding domestic component base. Printed-circuit-board (PCB) manufacturing is an essential building block for creating contemporary electronic equipment. According to the India Printed Circuit Board Association, roughly two-thirds of

India's PCB market is served through imports. India generates a meager 0.7 percent of the world PCB output.[7] According to a PCB industry expert, "it will be difficult for India to compete against volume producers in China and Taiwan[,] and even those in Vietnam, Thailand, and Malaysia, unless makers are prepared to spend more than [US]$100 million."[8]

In the strategically important telecom-equipment industry, PCBs and a variety of electronic components account for around 90 percent of the product cost. However, with the exception of cable harnesses and packaging, no such components are currently made in India.[9]

While the liberalization (i.e., the sequential abolition of central Indian government licensing and control) of telecom services has boosted the demand for telecom equipment this has not led to the development of a domestic Indian telecom manufacturing industry. Instead global telecom equipment vendors such as Alcatel, Ericsson, and, increasingly, Huawei and ZTE have been the primary beneficiaries.[10]

Consumer electronics, the largest segment of India's electronics market, is largely dominated by multinational corporations (MNCs)—especially Panasonic, Sony, LG, and Samsung. Over the last few years these companies have substantially decreased domestic Indian production and now rely overwhelmingly on imports from China.

Televisions (TVs) make up the largest segment of India's consumer-electronics market. With the transition to liquid-crystal-display (LCD) TVs, local production of TVs in India has virtually stopped. It is important to emphasize that Indian vendors rely little on domestic production—they source almost entirely from China.[11]

A particularly telling sign of the status of India's electronics industry is that India does not even show up in McKinsey's list of top-ten countries in the global value-added electronics industry.[12] FDI in India's electronics industry has been extremely low, even relative to other sectors—the industry ranks twenty-six out of sixty-four sectors in terms of the cumulative FDI received from April 2000 to April 2013.[13]

A Challenging Resource Environment

Only a few years ago India seemed to be well placed to mobilize the resources needed to unblock the barriers to investment in electronics manufacturing and to fast-track the industry's growth. In 2009 India's current-account deficit (CAD) was US$26 bn (in current US$) and India's GDP was growing at above 8 percent annually.

In 2013 the constraints on government policies are severe. The Indian government faces a difficult task in attempting to bring the CAD down from US$88.2 bn during FY2012 to US$70 bn in FY2013.[14] Affected by the high CAD, the rupee has declined sharply, reaching a new low of 68.85 rupees to a US$ in late-August 2013. While the Indian currency has since strengthened, the rising cost of crude-oil imports and electronics imports, as well as India's weak manufacturing capabilities (especially in electronics), is likely to continue to exert pressure on the CAD.

India now runs a CAD of about 5 percent of GDP and a record fiscal deficit approaching 10 percent of GDP if state governments' debt is added.[15] India faces intensifying economic-policy constraints. Analysis and policy recommendations in this study are offered while considering these constraints.

Overdue Change

Bold action is required to initiate a break with the anemic growth of electronics manufacturing as well as with outdated patterns of policy responses. Things need to change and they need to change soon. To identify realistic options for the development of India's electronics manufacturing industry it is imperative first to analyze how much electronics manufacturing in India lags behind its potential.

The first chapter of this study therefore explores where India stands compared to its main competitors. The analysis sheds light on global transformations in technology and the markets defining India's new manufacturing imperative; highlights a weak industrial-innovation capacity constraining productivity growth; and points specifically to the disconnect separating manufacturing from India's design capabilities.

INDIA'S NEW MANUFACTURING IMPERATIVE

The plight of India's electronics manufacturing industry is part of a broader challenge. As highlighted by the Planning Commission, "[t]he slow pace of growth in the manufacturing sector at this stage of India's development is not an acceptable outcome. . . . While the services sector has been growing fast, it alone cannot absorb the 250 million additional income-seekers that are expected to join the workforce in the next 15 years. *Unless manufacturing becomes an engine of growth, providing at least 100 million additional decent jobs, it will be difficult for India's growth to be inclusive.*"[16]

Significant gains are also required in India's manufacturing productivity: "[T]o increase exports as well as provide its internal market with domestically produced manufactured goods that compete with imports, India must manufacture a much larger volume of products at competitive costs and quality."[17]

India, unlike China and other earlier industrial latecomers from Asia, can no longer rely *exclusively* on "high-volume, low-cost" manufacturing as the main strategic option for expanding its manufacturing industry. This traditional manufacturing paradigm has ceased to be the only viable strategy for India and has important constraints on its future viability.

Vertical specialization through geographically dispersed global corporate networks of production and innovation has fragmented industrial manufacturing. Product-life cycles are being drastically reduced and mass customization based on "low-volume, high-value" production is gaining in importance relative to traditional forms of "high-volume, low-cost" manufacturing.[18]

Current challenges faced by the global electronic manufacturing service (EMS) industry illustrate the limits to "high-volume, low-cost" manufacturing. Falling PC sales and slowing smartphone sales are squeezing profit margins while growth in tablets (tablet computers) and servers used in giant data centers are insufficient to compensate for this loss. Global brand leaders in computing and mobile devices are all experimenting with new and unfamiliar products—big tablets, small tablets, hybrid notebook tablets, ultra notebooks, and wearable computing devices—resulting in a proliferation of new models shipping individually only in low volumes. Profit margins of EMS providers, such as Taiwan's Foxconn, are squeezed since companies must now spend time testing how best to make each new product and as massive investments are required in restructuring companies' product lines.

While "high-volume, low-cost" manufacturing remains important, this traditional form of electronics manufacturing is facing increasing pressure. New opportunities are opening up for "low-volume, high-value" strategies. Advanced manufacturing technologies—new materials, nanotechnology, and additive manufacturing (often called "3D printing" or "3DP")—are transforming the economics of industrial manufacturing and innovation and are challenging existing distributions of competitive advantages.[19] Companies and governments in the United States, the European Union, and Japan are all searching for ways to strengthen their capacity for "low-volume, high-value" manufacturing and related services.[20]

Additive manufacturing (3DP) is an emerging technology which creates objects by adding material one extremely thin cross-sectional layer at a time. Conceptually it is similar to creating a larger object by stacking LEGO® building blocks. Additive manufacturing differs fundamentally from earlier established approaches to machining or shaping manufacturing where material is subtracted from a larger piece of material ("subtractive" manufacturing).[21]

Additive manufacturing may well fundamentally change the economics of industrial manufacturing. Massive cost reductions in process technology become possible as 3DP reduces scrap, shortens production cycles, and increases flexibility in product design and development. Additive manufacturing will create a new geography of industrial manufacturing. Assembly lines and supply chains will be reduced or eliminated for many multi-component products today sourced from plants around the world and assembled in specialized assembly lines.

As a potentially disruptive technology, 3DP could have negative implications for the established "high-volume, low-cost" manufacturing model focused on export-led growth. As this new technology becomes price-competitive and widely deployed, the production of many goods may shift in the future back to consumer countries such as the United States. Such a change may lead to falling demand for imports from Asian emerging economies.

This poses a major challenge for "high-volume, low-cost" export production factories across Asia, especially for such factories in China. In the words of Richard A. D'Aveni, the transformation of industrial manufacturing through "this new technology will change again how the world leans."[22] Potentially this may reverse the last decades' transfer of wealth and jobs to Asia generated by the offshore outsourcing of manufacturing from developed countries.

Leading Asian exporting countries, however, are not sitting still. They are searching for ways to co-shape the development of transformative advanced manufacturing technologies. China, for example, is aggressively developing a complete laser-industry chain as a basis for manufacturing 3DP equipment. This evolving industry will address the manufacture or development of component crystals, electronics, lasers, accessories, and operating systems as well as R&D,

Advanced manufacturing technologies—new materials, nanotechnology, and additive manufacturing—are transforming industrial manufacturing and challenging existing distributions of competitive advantages

applications, and service. China also is very active in new materials, nanotechnology, advanced computing, and new energy technology.

From India's perspective it is important to emphasize that 3DP is one of the priority targets of China's innovation policy, as noted especially in China's "Strategic Emerging Industries" initiative. This includes China's creating medium- and long-term development strategies for 3DP, promoting the formulation of codes and standards, and increasing efforts to support 3DP technology development and commercialization through special fiscal and tax policies.[23]

Through such efforts, China today has the fourth-largest installed base of 3DP users with 8.7 percent of all industrial 3DP installations.[24] According to Luo Jun, chief executive officer of the Beijing-based Asian Manufacturing Association, revenues for the 3D printing industry in China are likely to reach 10 bn yuan (US$1.6 bn) within three years. A leading industry expert expects that China may become the biggest 3DP market within three to five years.[25]

Of particular interest for India's electronics industry is that the potentially most important application for 3DP is in producing parts and components for final products. This is already happening in China in the defense and aviation industries for "low-volume, high-value" components.[26]

As this study demonstrates, the lack of a vibrant domestic component industry is one of the most fundamental weaknesses of India's electronics industry. This raises the question whether India could catch up, through the use of 3DP manufacturing, to enable India-based manufacturers to domestically produce components which otherwise would have to be imported at high cost.

Given that 3DP manufacturing technology is still at an early stage, with many unresolved technical problems, there is no doubt that the costs and risks involved in such technology leapfrogging would be substantial.[27] It may, however, be worth exploring to what degree India's defense and aviation industries are suited to become early adopters of 3DP.

India's manufacturing imperatives will be shaped by newly emerging advanced manufacturing technologies. For India's electronics industry, this raises an important strategic challenge. Can Indian electronics manufacturing companies develop the necessary capabilities to substantially expand their presence in "high-volume, low-cost" manufacturing while, at the same time, pursuing a niche-market strategy focusing on higher-value products generating premium prices and enabling sufficiently large profit margins to support R&D investment? Failing to meet this challenge poses risks to both the industry and to India's goals for job creation—but the key factors driving the challenge are beyond India's control.

DISCONNECT BETWEEN MANUFACTURING & DESIGN CAPABILITIES

India seems, in principle, to be well qualified to address this new manufacturing challenge. Accumulated strengths in electronic systems and integrated-circuit design could provide the basis for such "low-volume, high-value" electronics manufacturing. Such strengths, however, are not evenly distributed.

While design skills in India are well developed for digital design, embedded software, and reference-board design they lag behind global standards in analog- and mixed-signal design.

According to the Twelfth Five-Year Plan Working Group on the Information Technology Sector, Indian electronic-design engineers "[lack] the breadth and the depth of experience, where breadth indicates the knowledge of all the aspects of a design flow and depth indicates an extensive knowledge of a particular aspect of the design flow."[28] Of particular concern is that India lacks sufficient capabilities in end-to-end chip design and, most importantly, in the critically important area of design-for-manufacturing. Systems management still continues to be driven by the overseas headquarters of MNCs.

India's thriving integrated-circuit design sector remains largely disconnected from the India market. Most of India's integrated-circuit design work is done for MNCs and these designs are then transferred to the headquarters of the MNCs where decisions are made on where to locate manufacturing—many times ending up in such locations as Shenzhen, China, rather than in India.

Deep integration of electronic-design capabilities into global R&D networks is paired in India with almost no integration into the domestic electronics manufacturing value chain. Major electronic-design automation tool providers for chip design have large facilities in India. But these facilities are almost entirely focused on export markets. Little of these capabilities are disseminated within India.

Indian manufacturing is not benefiting from its rich national pool of sophisticated IC design engineers. This creates a lack of engineering talent for product conceptualization and product management for the Indian market and for emerging markets. To reap the benefits of value-chain integration, India needs to devise a way to link IC design-and-development capabilities now trapped within MNC R&D laboratories back to India-based companies serving India's domestic markets. A related challenge is how to strengthen interactions and knowledge flows between the MNC R&D laboratories and India's university laboratories and public research laboratories. Currently these capabilities for innovation and production are fragmented where they need to be connected.

A FRAGMENTED INNOVATION SYSTEM

The result of the fragmented elements noted above is that India's weak innovation capacity constrains the growth of productivity. As emphasized by the Planning Commission,

> [t]he lacklustre growth of manufacturing . . . [is due] . . . to the low technological depth of the Indian manufacturing sector. In India R&D has not been sufficiently exploited and needs an overhaul in terms of its focus and its organization. Most Indian manufacturing firms appear to be stuck at the basic or intermediate level of technological capabilities. Creating conducive environments to increase business expenditure on R&D complemented by institutional measures around skill development, regulation and standardisation need to be key areas of emphasis.[29]

What are the sources of this fragmentation, specifically as they pertain to electronics manufacturing? The third chapter of this study will show that, for many domestic companies, inadequate size prevents economies of scale and scope while high costs of doing business and complex regulations constrain profit margins and hence investment in production and R&D.

Larger foreign original equipment manufacturers (OEMs) and EMSs typically conduct only final assembly, and are reluctant to invest in full-scale manufacturing and R&D, in India.

Industrial research outputs, in terms of patents, and non-technological innovation, in terms of trademarks, remain limited. No company in the interview sample has developed and successfully launched a radical breakthrough product innovation over the last three years. A few companies in the interview sample introduced new products but these were based on foreign technology and included mostly incremental adaptations or derivative product variations. Those companies are, however, reluctant to file for patents with the Indian Patent Office as its response time is much too slow for the needs of the fast-moving electronics industry. Most companies involved in the interviews had not filed patents.

Government policies, thus far, have had a mixed impact. Indian government funding of R&D accounts for more than two-thirds of the total funding sources,[30] but it has not been able to compensate for the weakness of industry R&D. Industry funding of R&D has steadily increased over the past 20 years, but remains less than a third of the total, whereas in the United States and China industry accounts for more than two-thirds of all R&D funding.

More broadly, compared to Asian and US competitors, the Indian government provides only limited support for industrial R&D. A significant portion of India's R&D focuses on support for its services sector—which accounts for about two-thirds of India's GDP. India's pharmaceutical industry also accounts for a sizable portion of its R&D while, with the exception of Bharat Electronics Limited (BEL, the public-sector enterprise), electronics manufacturing is practically absent from the national R&D scene.

The *Global Innovation Index 2013* provides ample evidence of India's weak industrial innovation capacity.[31] India ranks 66, out of 142 countries, with a total score of 36.2. China ranks 35, with a total score of 44.7. According to the Organisation for Economic Co-operation and Development (OECD), "India shows a relatively low capacity in science, technology and innovation . . . , in comparison to advanced OECD countries and emerging economies like China."[32]

According to the Battelle Institute, a primary source of international R&D data, India's gross expenditure on R&D was 0.85 percent of GDP in 2012 (compared to 1.6 percent in China), a percentage essentially unchanged since 2000.[33]

Figure 2 (below) shows that—while India is on par with France, the United Kingdom, and Russia in its total annual R&D expenditures—it substantially lags behind not only Germany and the United States but also China and Korea. For critical indicators, such as R&D intensity and the relative number of scientists and engineers, India's R&D system remains second tier—roughly at par with smaller countries such as Iran, Malaysia, South Africa, and Turkey.

India's gross expenditure on R&D was 0.85 percent of GDP in 2012 (compared to 1.6 percent in China), a percentage essentially unchanged since 2000

Figure 2.

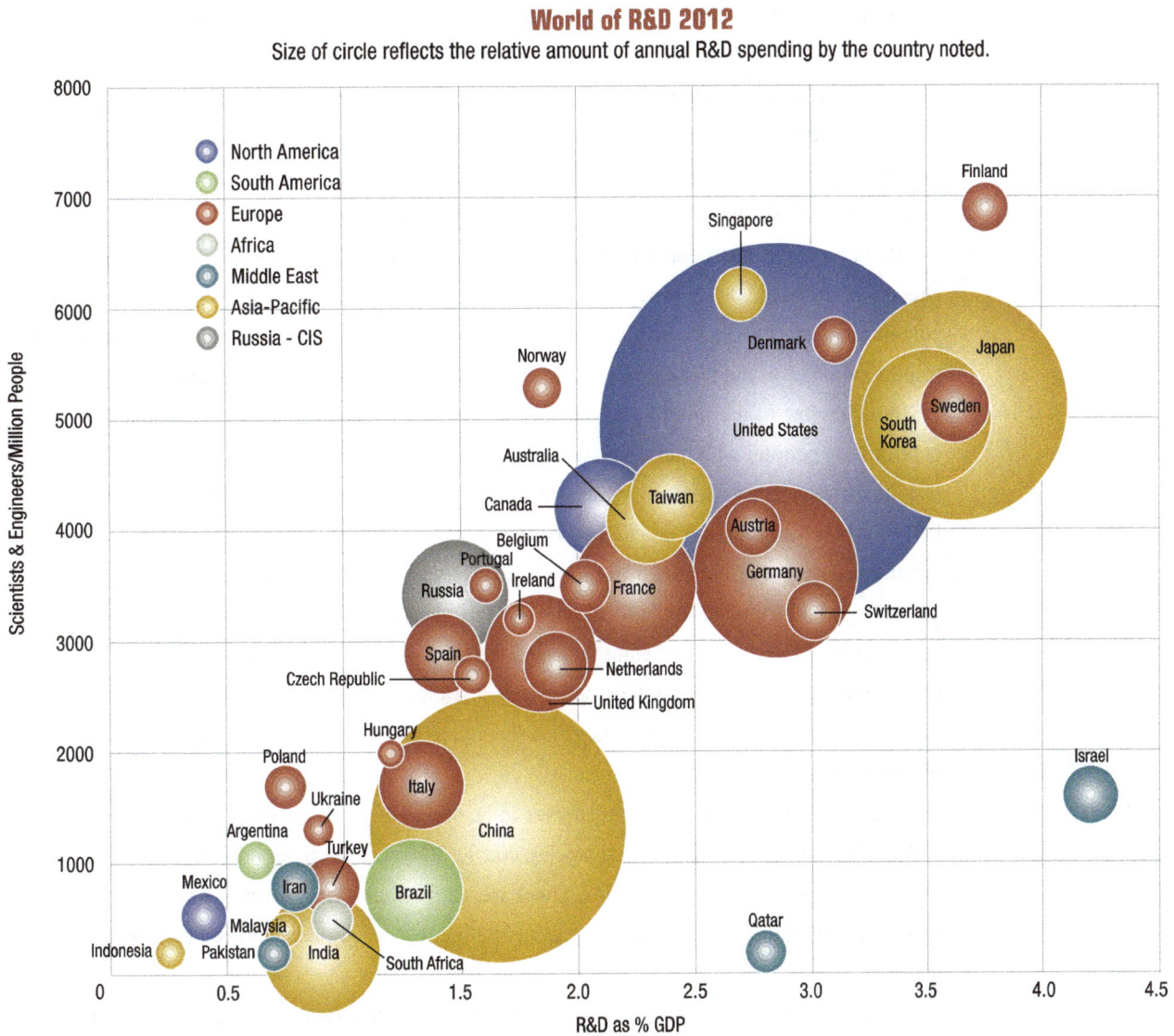

World of R&D 2012
Size of circle reflects the relative amount of annual R&D spending by the country noted.

Source: Battelle, *R&D Magazine*, International Monetary Fund, World Bank, CIA World Factbook, OECD

Adapted with permission from the 2014 Global R&D Funding Forecast, R&D magazine, December 2013, www.rdmag.com

India's industrial-innovation capacity today contrasts with its long tradition—with then-advanced technology development dating back to 2500 BCE.[34] For centuries India was the world's largest economy, producing a third of the global gross domestic product.[35] By the end of the seventeenth century it controlled a quarter of the world trade in textiles[36] and, as late as the eighteenth century, the British government dispatched observers to India to study innovations in steel, textiles, and medicine.

Today India's innovation system is characterized by persistent heterogeneity. One economy is the high-productivity, knowledge-based organized sector producing 41 percent of GDP but absorbing only 8.6 percent of the workforce.[37] The other, vastly larger, economy is the informal sector absorbing 91.4 percent of the workforce but contributing only 59 percent of GDP.

Both economies coexist but remain separated by a huge divide in productivity, capital intensity, and skills and move at different speeds. India's innovation system remains complex and fragile and it is difficult to predict its future development.

India's economic institutions, both public and private, were largely designed for a time before India was opened to the global economy. These institutions must be strengthened to cope with the requirements of transforming India into an internationally competitive industrial economy.[38] The task of modernizing India's economic institutions has only just begun.[39]

There is a deep fragmentation of India's innovation system with weak links between education, research, and industry.[40] With but few exceptions, India has a poor track record in commercializing ideas, discoveries, and inventions. Policies have emphasized self-reliance and techno-nationalism but neglected applied research and commercial-product development.

India has one of the smallest ratios of "scientists and engineers per million people" (137/million people) with "an estimated 25 percent shortage of engineers in the country."[41] This is quite different from China and Korea, which tie as the top producers of engineers in the world. Part of this skill shortage is due to the highly variable quality of India's higher education institutions, ranging from excellent to inadequate. Elite institutions such as the Indian Institutes of Technology (IIT) and the Indian Institutes of Management (IIM) cannot provide all the answers. All of India's IIT educate just eight thousand students—a tiny fraction of the country's population. All IIM, taken together, accept fewer students than Harvard Business School.[42] Of particular concern is the dearth of qualified faculty, as top graduates prefer to go abroad or work for the affiliates of global corporations.

India's higher-education system appears to be one of the weakest links in its innovation system. With tuition fees and self-financing courses, the cost of higher education is shifting from the government to private households. There are, additionally, widely recognized deficiencies in Indian primary education, which is damaged by high teacher absenteeism and high student drop-out rates.[43]

Recent research shows that "the persistently high illiteracy rate is falling, . . . and almost none of the new entrants to the urban labor force is illiterate but a majority of them do not complete school."[44] These shortfalls in general education severely limit the quality of those workers needed by India to translate innovations into competitive products and services.

A final element of concern is engineering and vocational training. Research for the Working Group on Information Technology Sector for the Twelfth Five-Year Plan documents pervasive general-skill gaps in India's electronics industry: "While electronics engineers lack skills and exposure to production processes, mechanical engineers lack sufficient exposure to electronics, but the industry/shop floor expects multiskilling in both these areas."[45]

The working group reports widespread skill bottlenecks including those in surface-mounted technologies (SMT), LCD technologies, semiconductors, nanotechnology, programmable logic controllers and robots, and quality-control practices and tools. Serious skill bottlenecks are also reported for basic manufacturing tasks such as precision welding and soldering, safety norms, meeting pollution-control laws, reading circuit diagrams and populating boards, and awareness of shop-floor concepts of electrostatic discharge (ESD). These skills are complex and take time to train and may therefore be precisely the skills neglected in attempts to reach exceedingly ambitious targets to train hundreds of millions of workers in just a decade. Such an expectation favors short, simple, courses fitting easily into government-procurement norms.[46]

Beyond electronics manufacturing, but of vital importance to it, India faces a challenge of historic proportions. Converting India's demographic surplus into a source of sustained economic growth will require India to repair its education system. Education and research must be linked together so that both foster the development of domestic industry. This is necessary to improve India's competitiveness and its innovative capabilities in industrial manufacturing.

Should this linkage not occur, India's electronics industry cannot count on a vibrant industrial-innovation system. The lack of such a system in turn constrains India's capacity for productivity-enhancing innovation.

Such is the context in which this study was shaped: a weak industry with a fragmented innovation system faces an industry going through rapid global change ending historical strategies for growth.

The next chapter turns to the parameters constraining potential policy responses.

Policy Parameters

THE CHALLENGE

Regulatory Reform and Industry Support Policies:
In The Context of Institutions, Markets, & Trade Agreements

This study examines restrictive regulations imposed by the central and state governments and their support policies—with a particular focus on implementation capacity—to identify ways to unblock the development of India's electronics manufacturing industry.

We define "regulations" broadly to include *laws* (such as subordinate legislation, bylaws, and amendments), *policies* implementing regulations; and *rules, technical standards, directives, guidelines,* and *administrative procedures* at the central, state, and local levels of Indian government.[47] Based on interviews sampling representatives of over forty-five companies, this study examines the impact on firm behavior of *fiscal regulations* (taxes, tax breaks, and other fiscal incentives), *trade-related regulations* (tariffs, technical barriers to trade, and customs clearance), *technical standards* and *certifications*, and *competition policies* (preferential market access and insolvency regulations) as well as *labor laws* and *regulations*.[48]

Subsequent chapters of this study will show that effective regulatory reform can help unblock barriers to investment and growth in electronics manufacturing through a pragmatic focus on incremental improvements. Such an approach would be in line with a 2008 report by Raghuram Rajan arguing that, rather than major politically controversial reforms, India should "take a hundred small steps that will collectively . . . [e]nsure inclusion, growth and stability by allowing players more freedom, even while strengthening the financial and regulatory infrastructure."[49]

Though necessary, however, effective regulatory reform will not be enough. To address the root causes of India's lagging performance, a longer-term and structural industrial-development agenda is also required. An important step in this direction is the National Policy on Electronics (NPE). This policy initiative seeks to improve India's international competitiveness through incentives for capability development, cluster formation, R&D, and technology transfer through FDI.

It is, though, important to first understand the institutional and structural constraints on this regulatory reform and needed support policies.

Recent research on trade liberalization in India and elsewhere finds the strength of a country's economic institutions conditions the success or failure of attempts at regulatory reform.[50] Regulatory reforms need to be appropriate to other elements of the economic environment such as the state of technology and the organization of credit and labor markets.[51] In line with this emphasis on domestic economic institutions, Raghuram Rajan argues in a recently published essay that India's economic growth has slowed because:

> India probably was not fully prepared for its rapid growth in the years before the global financial crisis . . . [S]trong growth tests economic institutions' capacity to cope, and India's were found lacking . . . [And] . . . because India's existing economic institutions could not cope with strong growth, its political checks and balances started kicking in to prevent further damage, and growth slowed . . . To revive growth in the short run, India must improve supply, which means shifting from consumption to investment. And it must do so by creating new, transparent institutions and processes, which would limit adverse political reaction.[52]

This second chapter of this study highlights how the legacy of the "License Raj" (India's post-1951 system of industrial licensing regulating and restricting the entry of new companies and the expansion of existing ones) shaped the transition from state-led mission-oriented planning to liberalization. Persistent restrictive regulations, however, continue to stifle private investment and innovation in India's electronics manufacturing industry (see "Domestic Institutions and the Legacy of the License Raj," below). While the story of the License Raj is well known in India it is important to understand its continuing legacy for the electronics industry—one of the last Indian industries to be liberalized.

It is also important to broaden the analysis beyond the role of domestic economic institutions. Equally important is an analysis of two interrelated international transformations:

- Constraints on India's industrial and innovation policies resulting from increasingly complex international trade agreements and their rules (see "International Trade Agreements," below).

- Entry barriers to Indian electronics manufacturing companies resulting from the prevalence of global oligopolies in the electronics industry (see "Global Oligopolies and Entry Barriers," below).

These concepts will be used in the following chapter to interpret the research findings on firm strategy and regulatory barriers. The framework also serves as a guide for examining, in the final chapter, implications for support policies (especially the NPE) and for evaluating recommendations received during extensive interviews with representatives of India's electronics industry.

Finally, this framework will also help identify a more general finding: There are diverse pathways to industrial upgrading in the electronics industry and India needs to develop its own approach (or portfolio of approaches). Given the constraints imposed by the above-noted three fundamental policy parameters, replicating the Japanese, Korean, or Taiwanese models of electronics industry development and upgrading is clearly not an option for today's India.

To revive growth, India must improve supply, which means shifting from consumption by creating new, transparent institutions and processes

DOMESTIC INSTITUTIONS & THE LEGACY OF THE LICENSE RAJ

From State-Led Mission-Oriented Planning to Liberalization

After independence, India's industrial policy had been shaped by the 1951 Industries (Development and Regulation) Act. This act introduced a system of industrial licensing regulating and restricting the entry of new companies and the expansion of existing ones and became known as the "License Raj."

It was believed that state control over industrial development through licensing would accelerate industrialization and economic growth and reduce regional disparities in income and wealth. Regulations encompassed all aspects of business from establishing a factory to starting a new product line.[53] Applications for industrial licenses were made to the Ministry of Industrial Development and then reviewed by an inter-ministerial licensing committee.

This rigid system of regulation prevented the development of a vibrant private manufacturing industry. Only a handful of the large business houses (such as the Tatas and Birlas) could afford to cope with the uncertainty resulting from unpredictable selection decisions and frequent delays of indeterminate length. The rigid regulations created perverse business models (e.g., the leading business houses routinely engaged in preemptive license applications as a means for stabilizing capacity and investment planning).

During the 1980s it became clear the License Raj development model was producing disastrous results. Rising external debt, exacerbated by the increase in oil prices caused by the Gulf War, resulted in a macroeconomic crisis. India was obliged to request a stand-by arrangement with the International Monetary Fund (IMF). In May 1991 a structural-adjustment agreement imposed by the IMF as a condition for financial aid became a powerful catalyst for the government to implement a far-reaching liberalization of the Indian economy. In a succession of industries, industrial licensing was increasingly abolished.

Liberalization Came Late to the Electronics Industry

The electronics industry was one of the last industries to be de-regulated. The main focus of the electronics industry was, until the 1980s, the so-called "strategic industries"—especially defense electronics. State-owned enterprises (SOEs) such as Bharat Electronics Limited (BEL) and the Electronic Corporation of India Ltd. (ECIL) dominated. The Department of Electronics (DOE) was in charge of approving not only a firm's entry into electronics but any changes in product line or increased output for a product already approved.[54] Common interests shared between the DOE and the SOEs "impeded the emergence of local private firms and delayed India's exploitation of new microprocessor-based technologies."[55]

Electronics industry licensing was abolished only in 1996 for consumer electronics. Significantly, the important aerospace- and defense-electronics sectors were excepted from this liberalization. As part of the dramatic shift to liberalization, tariff and non-tariff barriers were also slashed as India opened its economy to the outside world.

It was expected that the liberalization of industrial and trade policy would "encourage and assist Indian entrepreneurs to exploit and meet the emerging domestic and global opportunities and challenges. The bedrock of any package of measures must be to let the entrepreneurs make investment decisions on the basis of their own commercial judgment."[56]

Many dysfunctional regulations have remained in place and continue to stifle private investment and innovation in India's electronics industry. As will be analyzed in the third chapter, many companies believe that, while the quantity of licenses has declined, the cost of complying with those that remain has climbed. The fact that consumer electronics were de-licensed only in 1996 meant that when this industry segment was opened to the full force of free trade in 1997 it was still struggling to find its feet.

INTERNATIONAL TRADE AGREEMENTS

India's experience with trade liberalization through international trade agreements has had two sides. Some sectors—such as information technology (IT) services, car components, and generic pharmaceuticals—are seen to have benefitted from India's membership in the World Trade Organization (WTO). As far as electronics manufacturing is concerned however, this section will demonstrate that the gains from trade liberalization were overshadowed by substantial costs—especially those of stalled or declining domestic production.

This finding should not divert attention from this study's message that restrictive regulations and weak implementation of support policies are the principal constraints on investment in and growth of India's domestic electronics manufacturing. As emphasized by the Twelfth Five-Year Plan, as well as by Raghuram Rajan, India bears the primary responsibility for correcting its latecomer disadvantages in electronics and other industries.[57] As examined in detail in this study, the Indian government already seeks to fast track the development of India's electronics manufacturing industry through regulatory reform and industrial-support policies.

Nevertheless, it is important to emphasize that international trade agreements can act as *additional* constraints on India's domestic electronics manufacturing. The following analysis focuses on constraints on national industrial-support policies resulting from India's WTO membership and the impact of a proliferation of plurilateral trade agreements. The analysis will examine in detail the inverted tariff structure (finished products being duty free but their components not) favoring imports over domestic production and the resulting asymmetric distribution of gains from trade liberalization.

WTO-Related Parameters & "Plurilateral" Trade Agreements

India's NPE and other support policies and reform efforts directed at the electronics industry must consider important external WTO-related parameters. WTO membership obliges India to ensure "compliance" of its industrial and innovation policies with increasingly complex trade rules reflecting the evolution of the multilateral trading system. This constrains India's options for national support policies earlier available to Japan, Korea, and Taiwan.[58]

WTO membership constrains India's options for national support policies earlier available to Japan, Korea, and Taiwan

For instance, the Agreement on Trade Related Investment Measures (TRIMS) prohibits domestic regulations—such as those a country might apply to foreign investors—having trade-restrictive and distorting effects. Policies now banned, such as local-content requirements and trade-balancing rules, were earlier used to promote the interests of domestic industries.[59] As described in this study's closing chapter, the Indian government has had to delay its preferential market access (PMA) plan, a component of the NPE, in response to critique from the United States, the European Union, and Japan that this plan would not be in compliance with India's WTO obligations.

In addition, the WTO agreement on Trade Related Aspects of Intellectual Property Rights (TRIPS) sets down minimum standards for many forms of intellectual property (IP) protection to be provided by each WTO member to nationals of other WTO members.[60] TRIPS lays down in detail the procedures and remedies which must be available in each country so that rights holders may effectively enforce their rights.[61] Disputes between WTO members concerning respecting TRIPS obligations are subject to the WTO's dispute-settlement procedures.

According to its proponents, the TRIPS provisions seek to reduce distortions and impediments to international trade, promote effective and adequate protection of intellectual property rights, and ensure that measures and procedures to enforce intellectual property rights do not themselves become barriers to legitimate trade.[62] Critics argue that the current TRIPS provisions may impede both innovation and knowledge diffusion[63] and that they "commit countries to enforce the patents issued by other countries without any safeguards that . . . [those other] . . . countries are taking appropriate steps to guard against the issue of patents covering prior art, or trivial patents covering no art at all."[64]

Recent efforts by developed countries to push developing countries beyond their TRIPS commitments through "TRIPS-Plus" measures included in a growing number of bilateral and regional free trade agreements have raised concerns in developing countries. These concerns have been forcefully articulated especially by China and India. Concerns culminated when a group of mostly developed countries, led by the European Union and the United States, signed the controversial Anti-Counterfeiting Trade Agreement (ACTA) in October 2011.[65]

Overall, by introducing intellectual property law into the international trading system, the TRIPS agreement—and TRIPS-Plus measures—have fundamentally redefined the scope for national industrial and innovation policies.

Of similar impact is a "plurilateral" WTO agreement, the Government Procurement Agreement (GPA). As a "plurilateral" agreement the GPA is limited to WTO members which have specifically signed it or have subsequently acceded to it. Its present version was negotiated in parallel with the Uruguay Round of multilateral trade negotiations conducted within the framework of the General Agreement on Tariffs and Trade in 1994 and entered into force on January 1, 1996.

On December 15, 2011, negotiators agreed to re-negotiate the agreement, a political decision confirmed formally on March 30, 2012.[66] The agreement applies to all procurements for commodities, goods, and services if the maximum potential value of the contract will be in excess of US$552,000.[67]

The US government was an important driving force behind the GPA. According to the US Trade Representative, "the United States strongly encourages all WTO members to participate in this important agreement."[68] According to a document released by the US state of Massachusetts, the WTO-GPA "has been the United States' most effective negotiating tool for opening up opportunities for US suppliers to compete for foreign government contracts on a non-discriminatory basis. To date, the GPA has given US companies and their workers access to overseas procurement markets estimated to be worth more than [US]$200 billion annually."[69] India has observed the stipulations of the GPA since February 2010. It is noteworthy that, despite the fact that government procurement accounts for 25 to 30 percent of India's GDP, India does not have a central law on the subject of government procurement. As highlighted in the September 2011 Indian government Report of the Committee on Public Procurement,

Despite the fact that government procurement accounts for 25 to 30 percent of India's GDP, India does not have a central law on the subject of government procurement

> [a]t present, public procurement in India is governed by administrative rules and procedures which only attract departmental action in case of violation. These rules do not create any rights in favour of the public in general, and the potential suppliers, in particular. Nor do they provide for a fair and effective mechanism for dispute resolution, thus virtually denying any recourse against unfair and arbitrary decisions of the procuring entities. Another limitation of this arrangement is the absence of penal consequences for misrepresentation, cheating or fraud in public procurement, except under the normal penal codes which are inadequate for dealing with complex procurement matters.[70]

> Malpractices in procurement do not often carry any deterrent consequences and the associated lack of accountability enhances the potential for corruption. Departmental action against erring officials is rare, if not absent. Suppliers affected by malpractices have no recourse except through civil courts that are unable to offer any timely relief. As a result, public procurement does not inspire much public confidence.[71]

The government seems to be intent on using membership in the WTO-GPA as a catalyst for pushing through, against vested interests, national reforms outlined in the Public Procurement Bill.[72] However, there is a concern that the March 2012 revision of the text of the GPA is "watering down a number of flexibilities available to developing countries under the GPA 1994 by making Special & Differential Treatment . . . to developing countries, which was already a subject matter of negotiations in the GPA 1994, subject to more conditionalities and available only as transitional measures."[73]

The Information Technology Agreement (ITA)

For India's electronics industry, and for the NPE, the single most important plurilateral trade agreement is the Information Technology Agreement (ITA). India signed the ITA in 1997, one of the first developing countries to do so. India's rationale for joining the ITA was to attract inward FDI and facilitate the growth of its then-nascent IT services industry. ITA participation was also viewed as an important catalyst for further extending India's liberalization drive.

ITA participation and the resulting price reduction for information technology imports did, indeed, facilitate the expansion of India's IT services industry. At the same time, however,

India's participation in ITA has acted as an important barrier to the development of India's domestic electronics manufacturing industry. This is a trade-off which may or may not have been worthwhile.

Background. The ITA went into effect in April 1997 with twenty-nine WTO member countries and has since expanded to include seventy-eight member countries. It provided for zero tariffs for 217 electronics products.[74] The main product groups covered were computers, semiconductors, semiconductor manufacturing and test equipment, telecommunications equipment, software, and scientific instruments.[75] Not covered were consumer electronics products—including cathode-ray-tube (CRT) TV sets, video cameras, and photocopiers.

ITA-1 (the original ITA) enabled a substantial increase in the trade of the electronics products it covers. "Aggressive tariff liberalization facilitated growth in ITA trade from [US]$1.2 trillion to [US]$4.0 trillion . . . [in 2010]."[76] Unlike some other plurilateral trade agreements, such as the WTO-GPA (which allows exceptions by way of offsets, e.g., defense offsets), the ITA does not allow any exceptions for the products covered. The only relaxation of the requirements comes from identifying certain specified products as "sensitive" so that they may qualify for a phased-in implementation period. India has requested and received such staged extensions.[77]

Current negotiations to expand the product coverage, an "ITA-2," focus on product groups of particular interest to companies from developed nations. Such products include multi-component integrated circuits (MCOs),[78] medical devices, relay and industrial control equipment, optical media, and loudspeakers and handsets.[79] [80]

Conflicting perceptions of trade liberalization gains. Opinions differ on the distribution of trade-liberalization gains from ITA. A widely held perception in the United States is that "developing countries" benefited most from trade liberalization through ITA. For example, Ezell argued in 2012 that trade liberalization through ITA is likely to:

> benefit developing countries in three principal ways: 1) reducing tariffs on a broader range of ICT [information and communications technology] products encourages greater adoption of ICT products that play a key role in spurring economic growth; 2) lower prices realized by reducing tariffs on ICTs increases the productivity of all other industries in a developing economy; and 3) by lowering the price of a key input, the ITA has undergirded development of the burgeoning ICT software and services industries in many developing countries such as India, Indonesia, Malaysia, and the Philippines.[81]

However, the evidence provided to support this proposition is unconvincing to many. The argument neglects fundamental differences among ITA participants in their stage of development, in economic institutions, and in their resources and capabilities for manufacturing and innovation. Due to these structural differences, ITA participants differ in their capacity to reap these theoretical gains from trade liberalization.

Furthermore, as stated by the United States International Trade Commission (USITC) itself, "[t]he paucity of conclusive research on the impact of the ITA on global trade attests to the difficulties in empirically measuring the effects of the ITA and signals that . . . considerable discussion and analysis are still needed to determine the magnitude of the ITA's impact on IT trade and technology diffusion."[82]

A widely held perception in the United States is that "developing countries" benefited most from trade liberalization through ITA

Industry insiders and US officials have argued that leading US MNCs "benefit disproportionally" from ITA-enabled trade liberalization.[83] There is also some evidence that, for leading US vendors of ICT products, ITA provided significant benefits in terms of growing exports and expanding global production networks.

Semiconductors are an important product covered by ITA-1. While the US share in the worldwide market for semiconductors prior to 1997 was generally around 40 percent, since the signing of the ITA agreement the US share has moved up to around 50 percent.[84] From 2005 to 2009, semiconductors (on an aggregate basis) constituted the number one product export from the United States with exports totaling US$48 bn (US$10 bn more than automobile exports, the second-place export product).[85] In 2011, US semiconductor producers had global sales of US$152 bn, over one-half of the global semiconductor market.[86]

As for the impact on global production networks, the same research by the USITC found that ITA-1 boosted FDI by MNCs in China. This "had a major role in China's accelerating ITA exports, as multinational corporations sought to reduce costs by directly adding capacity in China. Once China joined the WTO, products exported from China were guaranteed MFN [most favored nation] access to other countries, providing strong incentives for multinational corporations to establish production and assembly operations in China."[87]

India's experience with ITA-1. It is important to emphasize that India joined the ITA from a position of weakness. The country was heavily reliant on electronics imports and its weak domestic electronics industry had only recently been liberalized.

India volunteered the largest tariff concessions of any ITA signatory, 66.4 percent based on pre-ITA inbound rates. This was far greater than the concessions of Thailand (30.9 percent) and Turkey (24.9 percent). India also stood out in average applied-tariff reductions. India's tariffs were reduced from a pre-ITA level of 36.3 percent as compared to China's average applied-tariff reductions which started from a level of 12.7 percent.[88] Overall, an industry weaker and more recently liberalized than any of its competitors was subject to a larger financial shock than any of its competitors.

This may be contrasted with China's approach. China joined the ITA only in 2003, six years after India. Unlike India, China entered the ITA from a position of strength. "China was . . . [already] . . . a leading manufacturer and trader of IT products prior to joining the ITA and deeply engaged in the global IT production chain even before tariff liberalization."[89]

When China joined the ITA in 2003, its per capita GDP (US$1,270) was three times higher than that of India's 1997 per capita GDP (US$427).[90] By 2003, China was already the third-largest exporter and the fourth-largest importer of ITA products. In 2004 China expanded its market share—becoming the world's largest exporter of ITA products. In 2005 China surpassed both the European Union and the United States to become the largest country in terms of overall ITA trade.[91] As China is far ahead in its electronics manufacturing industry, India is now an easy target for low-cost electronics imports from there.

The last decade has produced a significant acceleration of ITA imports into India. In 2000, 96 product lines were reduced to zero tariff and, in 2005, 121 product lines were reduced to zero tariff. Between 1997 and 2000 the growth rate of India's ITA imports was 18 percent annually—between 2001 and 2005 the growth rate of India's ITA imports increased to nearly 38 percent annually.[92]

Data from the Directorate General of Foreign Trade (DGFT) show that India's electronics imports under HS trade classification code 85 have grown faster than India's electronics consumption.[93] The import content of the raw material consumption of India's electronics industry has increased over the last seven years from 50 percent to 56 percent.[94] More recently, between FY2010–11 and FY2012–13, India's import of integrated circuits, the second largest electronics import category, has grown especially fast (82 percent). During the same period, India's imports also grew very fast for other electronic components such as capacitors (36 percent annually) and rectifiers and inductors (38 percent annually) as well as for consumer products such as video recorders and monitors (81 percent).[95]

Needless to say, trade deficits are not always, in principle, damaging to economic growth. Empirical research, in fact, points to the importance of imports in boosting productivity.[96] Yet in India's case, the local value-added for electronics manufacturing is only around 7 percent, while electronics imports account for almost two-thirds of consumption. Until there is a more substantial base of domestic production to benefit from such spillovers, positive productivity effects from rising imports appear unlikely.

The growing use of non-tariff barriers (NTBs) and technical barriers to trade (TBT) have further mitigated any positive effects of ITA-induced tariff reductions in target markets. The United States, the European Union, and Japan appear to be the main drivers behind the surge in NTBs and TBT. Of the total of 456 TBT notifications from 1995 to 2000 by all WTO members, developed countries have submitted 356 notifications, i.e. 78 percent of the total.[97]

There are few effective governance mechanisms in place to ensure that the surge of NTBs and TBT does not constrain access of Indian companies to the markets in the United States, the European Union, and Japan. These barriers may only be addressed by sophisticated institutions and governance capabilities, particularly for the development of standards (discussed below). Their absence creates an uneven playing field for India as it is still building such capabilities.

This would imply that, for developing countries such as India, the gains from trade liberalization through ITA may well be nullified through the surge of NTBs and TBT from developed countries.

India faces a double bind: While tariff reductions have led to a sharp decline in investments in domestic electronics manufacturing, exports from India face substantial NTBs and TBT in the United States, the European Union, and Japan.[98]

Participation in the ITA and the resulting inverted tariff structure appears to have had a negative impact on India's electronics manufacturing industry, one which must be taken into account in any future policy development.

The possible impact on India of ITA-2. In Geneva in July 2013, ITA members were negotiating a possible substantial expansion of the list of products covered by ITA. India decided not to join the Geneva ITA-2 negotiations.[99] It argues that developed countries have designed the parameters for a broadening of the scope and product coverage of the ITA (referred to as "ITA-2") and that this expanded list includes products where these countries, especially the United States, continue to lead by a wide margin.

Documents outlining the US negotiation strategy for ITA-2 support this view. In fact, a report of the USITC for the US Trade Representative identified five priority subsectors for ITA-2: medical devices; relay and industrial control equipment; optical media, including light emitting diodes (LEDs); loudspeakers and handsets; and, most importantly for India, multi-component integrated circuits (MCOs).

In 2011 estimated sales of MCOs accounted for between 1.5 and 3.0 percent of global semiconductor sales—an estimated US$1.2–US$2.4 bn.[100] [101] USITC selected these subsectors to illustrate "the potential for increased market access opportunities for USA firms as a result of ITA expansion."[102]

One strategy for India proposed by the Associated Chambers of Commerce and Industry of India (ASSOCHAM) National WTO Council[103] argues that merely resisting the expansion is unlikely to have positive effects—the Indian government must engage in a strategy of *co-shaping* the consolidated product list: "India needs to address the ITA expansion, weighing carefully its long-term as well as short-term objectives in a strategic manner rather than becoming overly influenced by *ad hoc* approaches and concerns."[104]

In this view, non-participation in ITA-2 negotiations would come at a heavy cost. Not only would India lose the opportunity for possibly co-shaping the content of the expanded ITA product list, non-participation might also act as a disincentive for existing FDI manufacturing projects to expand and upgrade their facilities. Additionally, while duty concessions achieved through FTAs require onerous rules-of-origin paperwork, the expanded ITA-2 is expected to provide long-term certainty and not require burdensome paperwork for industry.

The outcome of the July 2013 Geneva ITA-2 negotiations, however, raises doubts whether this strategy could succeed. In these negotiations China tried to implement precisely such a co-shaping strategy without encouraging results. China presented a list calling for the removal of 106 products rather than requesting an extended implementation period ("staging") for these products. Pressured to shorten their initial list, on July 17 China reduced it to roughly ninety products—but retained two product groups which were among the US priorities for ITA-2: MCOs and medical devices.

The following quote from a detailed report in the newsletter *Inside U.S. Trade* summarizes the ITA core group's response to the China's revised sensitivities list:

> Following that, the Canadian mission—which was organizing the meetings in Geneva— sent out a notice stating that talks previously scheduled for July 18 would not take place, on the basis of the earlier agreement at the ambassador level that talks could not advance without China producing a more "credible" list.
>
> Exactly what constitutes a credible list is something that no member has clearly defined, sources said. But one source said that the chief drivers of the ITA expansion initiative—the

U.S., Japan, and the EU—are clearly targeting a total expansion including about 200 items. That would require China to at least halve its current list of sensitivities.

China was not supportive of suspending the negotiations, and it is unclear whether it will really be able to back off its initial position to that extent. But while some sources charged that China's long list indicated a lack of coordinated domestic consultation—given what they claimed were the potential benefits China could reap from eliminating some of the tariffs it has asked to exclude—others said Beijing's position is more nuanced. It may be taking the stance it has because it has industrial policy goals in mind, one source posited.[105] [106]

What matters from India's perspective is that, if ITA-2 will indeed broaden the product list to include multi-component semiconductors (MCOs) and medical equipment, this would prevent India from producing these critical products. An additional aspect of China's ITA-2 negotiation approach of interest to India's NPE is that "China also listed some products already covered by the ITA such as printers and monitors, which has confused other negotiators."[107] These are in fact product groups where India has some, albeit limited, production capacity.

If ITA-2 would broaden the product list to include multi-component semiconductors and medical equipment, this will prevent India from producing these critical products

Regional and Mega-Regional Trade Agreements

The inverted tariff impact of ITA is further amplified by various free trade agreements (FTAs) and preferential trade arrangements (PTAs)[108] signed by India.[109] One of the most significant is the ASEAN (Association of Southeast Asian Nations)-India Free Trade Agreement. This agreement is expected to eliminate tariffs for about four thousand products (including electronics, chemicals, machinery, and textiles), with 80 percent of existing tariffs to be reduced by December 2013 and the remaining 20 percent of existing tariffs to be reduced by December 2016.[110]

The impact of this agreement on "electrical and electronic equipment" (a proxy for the electronics industry) is already being felt.[111] As research on the sectoral impact of the FTA states, "[T]here are hardly any immediate benefits for Indian producers as average percentage tariff drops in Malaysia, Indonesia, and Thailand's normal track products are much lower than India's. Further, the ASEAN-5 (Indonesia, Malaysia, the Philippines, Singapore, and Thailand) economies are leading exporters of light manufacturing products . . . [including electronics]. . . . India will also be competing with China and South Korea in the ASEAN market, which already have FTAs with ASEAN. Thus Indian SMEs (small and medium enterprises) will find it difficult to hold their own against these countries in such sectors."[112] As long as India lacks a vibrant domestic electronics industry, India's gains from the ASEAN-India FTA will thus be limited to its IT services industry (which, by some reports, is being out-competed by the IT services industry in the Philippines).

Important longer-term challenges may result from agreements India does not sign but which affect India's potential markets. These are the emerging so-called "mega-regional agreements," i.e., "by-invitation-only" arrangements such as the Trans-Pacific Partnership Agreement (TPP) and the Transatlantic Trade and Investment Agreement (TTIA).

Some observers are concerned that "developing countries will be excluded from market share in the signatory regions. Also, since these mega-regionals are being negotiated outside the scope of the multilateral trading system, developing countries are prevented from negotiating the rules that will set standards for the trading system as a whole."[113]

The government of India as well as those of Brazil and China have expressed concern that, unlike in multilateral negotiations, the United States has more political and economic leverage over other parties in the TPP negotiations. This is a particular cause for concern with regard to intellectual property (IP) provisions. Recent research indicates that IP provisions proposed for the TPP are likely to be even more restrictive than those in ACTA.[114] Perhaps the primary criticism of the TPP is that it sets US intellectual property laws as the "norm" for all members. The TPP patent provision has been criticized primarily for its impacts on pharmaceuticals and medical devices but will likely also impact software and other information and communication technologies. "In particular, the TPP removes the requirement that an inventor disclose the 'best mode' of the invention, thus creating the possibility of inventors 'retaining the best for [themselves].'"[115]

Other observers highlight possible negative implications for developing countries as these countries, by definition, do not have developed countries' institutional capabilities or standards.

As noted above, this reduces the theoretical level playing field (the motivation cited by the advocates of such agreements) to, in practice, a highly uneven playing field in which superior standards-setting and other governance capabilities offer substantial advantages. Another structural feature of the electronics industry, which makes it depart in important ways from theoretical models of gains from trade, is that it is not freely competitive but is oligopolistic in almost all its segments.

GLOBAL OLIGOPOLIES & ENTRY BARRIERS

Global oligopolies have proliferated in high-tech industries. A widely recognized example would be the aerospace industry: Two suppliers dominate the manufacture of large commercial aircraft,[116] three suppliers dominate the market for jet engines, two suppliers dominate the market for brakes, and three suppliers dominate the market for tires.

A limited number of MNCs dominate in important market segments with the result that Indian companies are confronted with substantial entry barriers

The same pattern holds for the global electronics industry. A limited number of MNCs dominate in important market segments with the result that Indian companies are confronted with substantial entry barriers. Support policies for the development of India's electronics manufacturing industry therefore need to be informed by a deep understanding of these oligopolies and the entry barriers they create.

Technology-Centered Competition is Intensifying

The electronics industry is unrivalled in its degree of globalization. A defining characteristic is that competition is centered on the increasingly demanding performance features for electronic systems. Tablets, laptops, smartphones, and mobile base stations all need to become lighter, thinner, shorter, smaller, faster, and cheaper as well as adding more functions and using less power. To cope with these demanding performance requirements, engineers have pushed modular design and system integration. The result is that major building blocks of mobile handsets, as one example, are now integrated onto chips.[117]

Design teams must cope with this accelerating pace of change. Essential performance features are expected to double every two years, time to market is critical, and product-life cycles are rapidly shrinking to a few months. Only those companies that succeed in bringing new products to the relevant markets ahead of their competitors will thrive.

The root cause of these increasingly demanding requirements is the emergence of a "winner-takes-all" competition model, first described by Intel's Andy Grove.[118] In the fast-moving electronics industry, success or failure is defined by speed-to-market and return-on-investment and every business function, including R&D, is measured by these criteria.

The examples of Samsung Electronics and Apple illustrate to what degree extreme profit expectations have come to dominate investors' decisions. Despite an estimated 47 percent year-on-year rise of earnings (to almost US$10 bn per quarter) record, Samsung Electronics, at one time, had lost 17 percent of its market capitalization since the beginning of 2013—as its earnings rise fell short of analysts' expectations.[119] Apple's shares have fallen 40 percent from their peak in September 2012 despite the iPhone5 breaking unit-sales records.[120] Even global leaders such as Samsung and Apple are finding that generating record results is insufficient to satisfy investors expecting even greater growth.

This results in intense price rivalry among industry leaders, which is further fueled by the growing threat from lower-cost Chinese brands. To prevail, industry leaders must use their technological superiority to cut costs even further and to erect new barriers to entry. Intensifying price competition thus combines with intensifying technology-centered competition.

The Spread of Global Production and Innovation Networks

This intense competition has provoked fundamental changes in business organizations. To mobilize all the diverse resources, capabilities, and repositories of knowledge on time and at lowest cost, global corporations have responded with a progressive modularization of all stages of the value chain and its dispersal across boundaries of companies, countries, and sectors through multi-layered corporate networks of production and innovation.[121]

The extreme complexity of these global networks is difficult to fathom. According to Peter Marsh, the *Financial Times'* manufacturing editor, "[e]very day 30m tonnes of materials valued at roughly [US]$80 billion are shifted around the world in the process of creating some 1 billion types of finished products."[122]

While the proliferation of global production networks goes back to the late 1970s, a more recent development is the rapid expansion of global innovation networks (GINs) driven by the relentless slicing and dicing of engineering, product development, and research.[123] Empirical research documents that this has further increased the complexity of global corporate networks. GINs now involve multiple actors and companies differing substantially in size, business model, market power, and nationality of ownership. This has given rise to a variety of networking strategies and network architectures.[124]

While the proliferation of global production networks goes back to the late 1970s, a more recent development is the rapid expansion of global innovation networks

Flagship companies, those controlling key resources and core technologies and shaping these networks, are still overwhelmingly from the United States, the European Union, and Japan. However, there are also now network flagship companies from emerging economies—especially from Korea, Taiwan, and, more recently, China.

As the most prominent example, Samsung Electronics today has eight regional headquarters across the globe. Its production network covers nine plants in Korea plus twenty-seven elsewhere in Asia and across Europe and North America. Taiwan's Foxconn has thirteen factories (or, rather, gigantic factory cities) in China and a growing number of factories in Japan, Malaysia, Brazil, and Mexico as well as in Hungary, Slovakia, and the Czech Republic. Huawei's global innovation network now includes, in addition to six R&D centers in China, five major overseas R&D centers in the United States, Sweden, Russia, and the United Kingdom (as part of British Telecom's list of eight preferred suppliers for the overhaul of its fixed-line phone network).[125]

Late Entry Into Global Oligopolies Requires Extraordinary Efforts

In economic theory, markets are oligopolies when they are "dominated by a few sellers at least several of which are large enough relative to the total market to be able to influence the market price."[126]

While some oligopolies may lead to price distortions this is not the only effect of oligopolistic market structures. As highlighted by Joseph Farrell and Carl Shapiro, "[i]mplicit in the market structure we call 'oligopoly' is the presence of some important barriers to entry. Companies that own certain crucial assets are the incumbents, and others are at best potential competitors. These assets might take the form of intellectual property: patent rights to production technology, licenses to use such technology, or industrial know-how."[127]

As the examples of Samsung and Apple show, intense competition may occur among oligopolists. Nevertheless their ownership of crucial assets allows them to establish barriers to the entry of potential competitors. The nature of these barriers may differ depending on specific characteristics of the relevant industry sectors and market segments. These barriers may result from pricing—but oligopolists can also establish and raise entry barriers not related to price, drawing on a superior capacity to define technology trajectories, control intellectual property rights and brands, and shape critical technical standards and their enforcement rules.

Entry barriers confronting latecomers in the electronics manufacturing industry, such as Indian companies, result from a combination of superior assets and capabilities that global oligopolists were able to develop based upon their dominant market positions. Price-setting is one such capability and it is of critical importance in major segments of the electronics industry.

From India's perspective it is important to highlight the systemic nature of these entry barriers. Oligopolists can set lower prices, not only because they can source the relevant products from low-cost production sites through their global production and innovation networks, but also because of their control over leading-edge technology and their superior innovation capacity.[128]

The concentration data for key segments of the global electronics market, presented below in the section headed "Evidence—Tight Global Oligopolies in Important Electronics Market Segments," clearly demonstrate that global oligopolies have been established across the electronics value chain. While late entry can never be completely excluded, successful entry into

Oligopolists can set lower prices because they can source the relevant products from low-cost production sites, but also because of their superior innovation capacity these markets would require extraordinary efforts by Indian firms to develop superior business models and new technologies. Both the Indian government and the private sector would need to join forces and develop decisively longer-term industrial-development strategies combining smart regulatory reform and structural support for these industries.[129]

Late entry into highly dominated developed markets is possible and has been done by others (indeed, Korea's Samsung was once an industry entrant). Chinese companies designing integrated circuits for smartphones and tablets provide more-recent examples. In the smartphones market, Chinese chip makers Spreadtrum and RDA Microelectronics have gained significant market share by undercutting both Taiwan's MediaTek (a leader in the low- and mid-level handset market) and Qualcomm (the US-based dominant chip designer for high-end phones).[130] In the tablet market, Chinese chip-design companies (led by Fuzhou Rockchip Electronics and Allwinner) have captured 37 percent of the market (with their share still rising) for the key processor chips in non-iPad tablets.[131]

In the tablet-applications processor market, Chinese and Taiwanese tablet-chip companies together captured a 29 percent share of market volume in Q1 2013. While Apple and Samsung together still command a 50 percent global market, thanks to their in-house customers, it is nevertheless clear that the once-tight global oligopoly is under pressure. One indicator is that the entry of Chinese companies competing with lower-cost processors has accelerated the significant decline in the average selling prices of tablets from US$522 in Q1 2012 to US$461 in Q1 2013.[132]

It is important to emphasize, however, that such successful latecomer entry into a global oligopoly was only possible as a result of China's extensive long-term industrial-development strategy. Since the 1990s China has provided its electronics industry with substantial and sustained support policies ranging from cost subsidies to R&D funding. The semiconductor industry has been one of the priority targets of China's indigenous innovation policy. An important objective is to create a group of globally competitive semiconductor companies which will develop into global leaders in market share, manufacturing excellence, and innovation capacity.[133]

Evidence—Tight Global Oligopolies in Important Electronics Market Segments

Empirical research on global oligopolies has focused on concentration ratios (CR) within a given industry. The most-commonly noted concentration ratios are CR4 and CR8, respectively denoting the market share of the largest four and eight companies. Concentration ratios are usually used to show the extent of market control of the largest companies in the industry and to illustrate the degree to which an industry is oligopolistic. According to J.M. Blair, oligopoly begins when the four largest companies control more than 25 percent of overall market. Between 25 and 50 percent this oligopoly is loose and unstable but above 50 percent it becomes firm and clearly established.[134]

Wintel–still the predominant platform standard for PCs. Control over platform standards determines who, within the electronics industry, can shape technology trajectories and markets. The PC industry, for instance, has been dominated by two companies—Microsoft and Intel— which, together, have tightly controlled the "Windows programs" (the software operating systems for most personal computers) and the "Intel architecture" (the rules governing how software interacts with the processor on which it runs). More than 80 percent of PCs still run on the "Wintel" platform standard.[135]

In Q1 2013, the four leading PC vendors had a global market share of 49.1 percent[136] while the five leaders control almost 56 percent.[137] More recent data from IDC show an even higher degree of oligopolization[138]—the four leading PC companies account for 53.5 percent of the global market while the five leaders control a 59.6 percent market share. Today's global PC market has become a firm and well-established oligopoly. The irony is that this happens when the industry is widely considered to be in decline, given the growth of tablets and smartphones.

Smartphones. With the decline of PC sales relative to the increase in sales of mobile devices,[139] some observers are predicting a transition to a "multi-polar" world where "the market will be fought over by eight or nine more or less vertically integrated giants." In this view, Oracle, Cisco, and IBM are expected to vie for corporate customers while Apple and Google will control the markets for individual consumers.[140]

Instead we are witnessing the emergence of a new global oligopoly for mobile devices. In 2012, the four leading operating systems accounted for 94 percent of worldwide mobile-device shipments. Google's Android system alone controlled 68 percent. Projections for 2017 expect an even tighter global oligopoly with the four leading operating systems accounting for 99 percent of worldwide smartphone shipments.[141]

In Q1 2013, the four largest smartphone vendors had a global market share of just less than 60 percent, indicating a firm and well-established oligopoly.[142] Markets for mobile devices are now controlled by two dominant companies. Apple and Samsung control the majority of profit and a growing portion of sales. Apple has the advantage of completely controlling its hardware and software and Samsung has the advantage of manufacturing many of its key components. Only these two oligopolists have the size and deep pockets necessary to be able to sell across multiple product lines.

Hard-disk drives. An even higher degree of oligopolization can be found in hard-disk drives. Some would argue that the hard-disk drive is a dying industry as the form of storage is used by struggling PC makers while tablets and smartphones use solid memory rather than disk drives. With the projected dissemination of distributed computing through the "cloud," however, huge centralized data centers will require massive storage capabilities—thus supporting disk-drive demand for an extended period. In Q1 2013 the largest three disk-drive manufacturers[143] had a combined market share of almost 100 percent—clearly an exceptionally tight oligopoly.[144] Only ten years ago the market structure was very different.[145] As Figure 3 (below) shows, the hard-disk drive industry represents a case of rapid global oligopolization.

The four leading PC companies account for 53.5 percent of the global market while the five leaders control a 59.6 percent market share

Figure 3. Hard Disk Drive — The Path to Tight Oligopoly

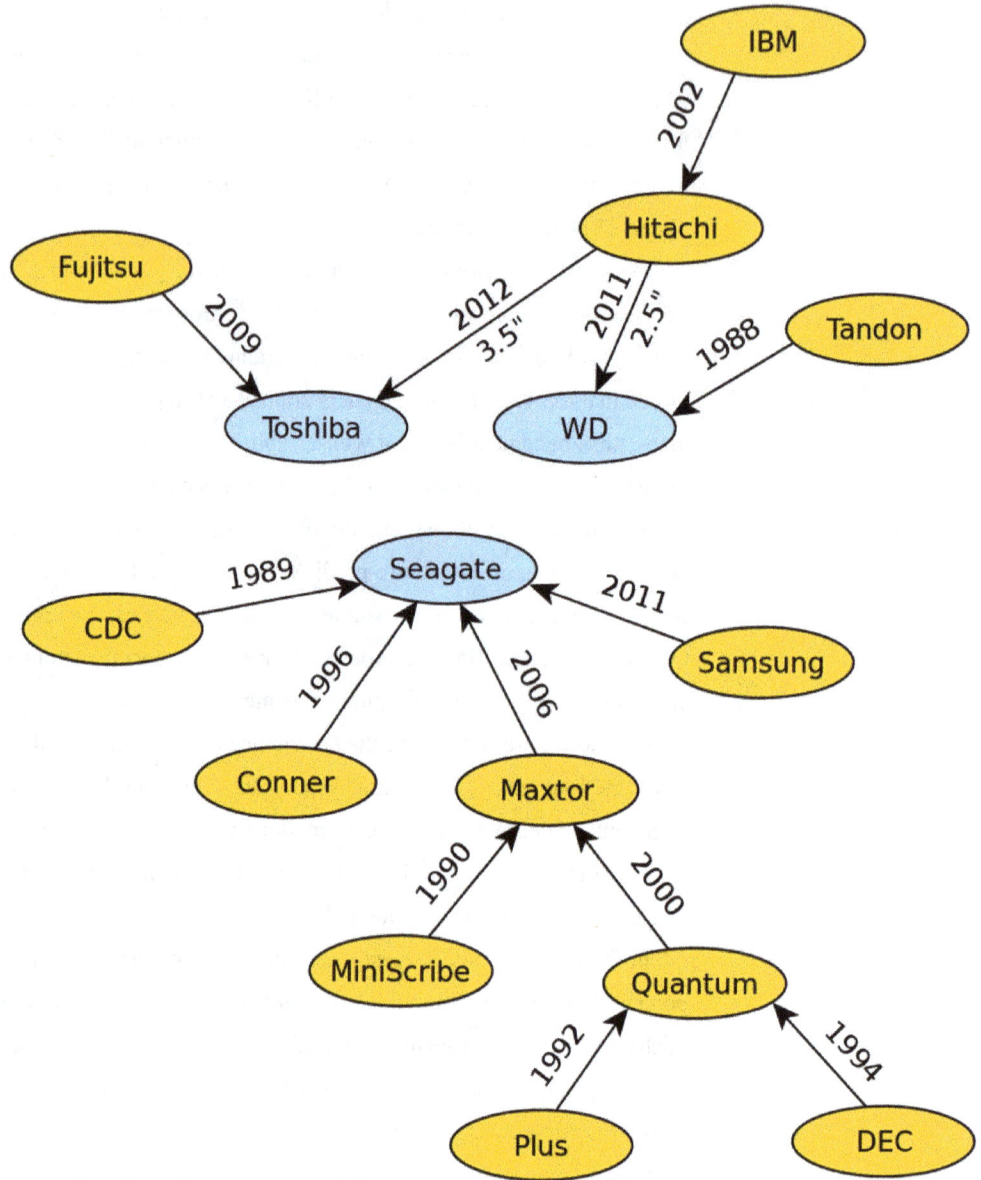

Source: http://en.wikipedia.org/wiki/File:Diagram_of_Hard_Disk_Drive_Manufacturer_Consolidation.svg. This is a file from the Wikimedia Commons, a freely licensed media file repository

Set-top boxes. Of particular interest for India's efforts to develop its domestic electronics manufacturing industry is the emerging oligopolistic market structure for key market segments of the set-top box (STB) industry. STBs are one of the identified priority products of India's NPE.[146]

This market is dominated by five companies: Pace, Motorola, Technicolor, Scientific Altanta/Cisco, and Humax.[147] In light of the segmentation of the STB market it is important to highlight the dominant market positions of leading companies in some of these market segments:

- Pace barely overtook Motorola in overall STB revenue in Q2 2011, attributed to the fact that Pace offers solutions across all market segments (telecom, cable, and satellite operators).

Five important electronics markets in India are dominated by a handful of MNCs

- Cisco holds a commanding lead in both revenue and units for Internet Protocol STBs,[148] Motorola leads in cable STB revenue, and Skyworth Digital dominates in cable STB unit shipments.[149]

Flat-panel televisions. The four leading companies (Samsung, LG, Sony, and TCL), have a combined market share of almost 47 percent, just less than a firm oligopoly.[150] However, the two leaders (Samsung and LG) together account for a third of the global market and their combined share in leading-edge seventh- and eighth-generation flat panel televisions keeps rising.

India's Electronics Market is as Oligopolized as the Global Industry

This global market structure is mirrored in India. According to data collected by the Department of Electronics and Information Technology (DEITy), five important electronics markets in India are dominated by a handful of MNCs (Table 2). For PCs, four MNCs controlled 57 percent of the Indian market in Q1 2013.[151] This constitutes a tight and established oligopoly. For smartphones, three MNCs accounted for 65 percent of unit shipments in Q1 2013 in India—with Samsung far in the lead with a share of 43 percent.[152]

Table 2.

Market	Companies
PCs	HP, IBM, Apple, Acer, Lenovo, Dell
Telecom equipment	Alcatel-Lucent, Ericsson, Nokia, Huawei, Cisco, Juniper
Flat-panel televisions	Samsung, LG, Sharp, Panasonic, Sony, Toshiba, Hitachi
Semiconductors	Intel, Samsung, Texas Instruments, Toshiba, AMD, ST Microelectronics, Analog Devices
Mobile handsets	Apple, Samsung, Nokia, RIM, LG

A few oligopolists also dominate India's huge and rapidly growing market for telecommunications equipment. In 2011, four companies (Nokia Siemens Networks,[153] Ericsson, Huawei, and ZTE) accounted for over 90 percent of the Indian market.[154] Huawei and ZTE are successfully attacking the others and together hold a 35 percent share in FY2012.[155]

Only the market for mobile telecom towers has a substantial Indian presence, with Indus Towers leading with 32 percent, followed by BSNL (15 percent), Reliance Infratel (15 percent), Viom Networks (11 percent), Bharti Infratel (10 percent), and GTL Infrastructure (10 percent).[156] These telecom-tower companies are essentially assemblers (engineering, procurement, and construction companies) which acquire a piece of land, erect towers (through vendors), and then rent out these towers to operators. Very few electronics components go into tower manufacturing[157] and the base station is typically provided by the operator.

Market Control Without Domestic Manufacturing

The multinational oligopolists described above dominate the Indian market *without* engaging in substantial domestic manufacturing in India (either directly or through EMSs), except for low-value-added final assembly. They rely on their extended global production networks to source the relevant products for the Indian market from other production sites, primarily in China.

It is important to note that MNCs do not consider just the basic cost structure in making these sourcing decisions but what they would have to invest in creating a whole range of sophisticated capabilities for rapid, low-cost, scaling up of large production lines for complex products. Those capabilities are difficult to acquire. Earlier research on Korea, Taiwan, and, more recently, China demonstrates that developing these sophisticated scaling-up capabilities carries a much larger cost than the physical investment in plant and equipment.[158]

So, in addition to China's substantial cost advantages, MNCs there benefit from the accumulated capabilities found in China for rapid and customized scaling up.[159] There are signs that such capabilities allow Chinese companies to charge even higher prices than Indian competitors while still winning orders. Witness the following statement of Hitech Magnetics, an Indian component supplier: "We have recently lost out three products that we were supplying to ABB over the last five years. MNCs now have global sourcing and they are asking us to be 15 percent lower than Chinese cost, only then they will source from us."

India's electronics manufacturing industry has not yet developed such scaling-up capabilities. In combination with India's substantial cost disadvantages relative to China, this explains why the MNCs dominating India's electronics markets have little incentive to invest in an expansion of local production in India. At the same time, these very same MNCs can use their power as global oligopolists to erect high entry barriers for Indian companies in case they would seek to enter or re-enter the industry.

Where local Indian companies seek to compete with MNCs for the India market they follow the same pattern of sourcing their products from offshore production sites in China. Support policies designed as part of India's NPE are intended to incentivize domestic companies to invest in domestic production.

Should Indian companies start domestic production these companies would face difficulties even within their own national market in challenging the dominance of MNCs. As oligopolists, MNCs can establish high entry barriers, drawing on their superior economies of scale and scope, long investment in low-cost and scalable production, mastery of expensive and leading-edge technology, and control over rich patent portfolios. Trade rules, as described above and including the inverted tariff structure, mean that the capabilities of MNCs create an entrenched barrier for any Indian companies considering attempting entry into the Indian market.

India thus faces a vicious cycle in its efforts to develop a domestic electronics manufacturing industry. As long as the industry is shaped by oligopolistic competition and an inverted tariff structure, neither MNCs nor Indian companies have much incentive to invest in substantial domestic manufacturing in India. Oligopolistic control gives rise to a "commoditization" of electronics products across the globe, imposing substantial constraints on local innovation efforts which seek to address specific needs of India's domestic market through "frugal innovation."

This raises important questions which will be addressed in the fourth chapter of the study: What policies, if any, would enhance the chances of Indian companies to overcome these entry barriers? How would it be possible for Indian companies to develop products addressing local needs and requirements including those specific to local languages, cultural needs, operating

"MNCs now have global sourcing and they are asking us to be 15 percent lower than Chinese cost, only then they will source from us"

conditions, etc.? And does the success of China's low-end "budget" smartphones indicate that frugal innovation might still be possible in global oligopolistic markets?

CONCLUSIONS

International trade agreements (such as the ITA) should, in principle, strengthen the multilateral trading system by reducing barriers to trade which have not been adequately addressed in multilateral trade negotiations. From a global welfare perspective such trade expansion should reinforce the diffusion of innovation.[160]

This would, however, require a greater balance in the distribution of gains from the ITA. As India's experience with ITA demonstrates, countries at different stages of development and with different economic institutions may find it difficult to reap equal gains. It is necessary to acknowledge the asymmetric effects that plurilateral agreements such as ITA may have on cost structures and capabilities of different participants. One way to reduce these asymmetries would be, in line with Mari Pangestu's suggestion, that international trade agreements "promote economic and technical cooperation recognising the different stages of development of participants. Special and differential treatment can be justified in circumstances where participants face challenges in benefitting from an increase in trade."[161]

These trade agreements intersect with a market which, since the agreements were signed, has become more and more oligopolized. This results in enormous barriers to entry faced by domestic companies not just in India but also in overseas markets. Domestic companies face powerful MNCs which combine developed-world skills and intellectual property with the capabilities for rapid, low-cost scaling up (as is available in China). Both are the result of decades of investment. Even if companies in India or elsewhere could *in theory* be competitive, in practice they will sustain enormous losses before acquiring the necessary management and technological capabilities.

None of those constraints are impossible to overcome. China is itself a demonstration that it is possible—both in its original acquisition of these capabilities and in the rise of Chinese companies to challenge the global oligopolies—particularly in mid- and low-priced segments. For China to successfully do so, however, it has had to engage in highly interventionist, long-term, support policies. These have included substantial subsidies of a range of inputs as well as creating an exceptional operating environment for the MNCs to invest in and for both MNCs and domestic companies to import and export.

India could, in principle, also achieve this. With electronic-goods penetration in China nearing saturation over the next few years in a range of markets, India will be one of the largest, if not the largest, growth market for electronics. Given the premium in this industry on time-to-market and customization, this should create an inbuilt advantage for domestic production in India—certainly when compared with any other developing market.

To achieve this, though, India will have to overcome the substantial latecomer disadvantages described above—and do this in a context where India's influence on key external parameters, in particular market structure and trade rules, is limited to non-existent. The responsibility for

It is necessary to acknowledge the asymmetric effects that plurilateral agreements such as ITA may have on cost structures and capabilities of different participants

overcoming this challenge thus lies domestically in the alignment and implementation of domestic policies and the improvement of the domestic business environment.

Only such a concerted reform and industry support effort will have a chance to unblock the barriers to investment and growth in the electronics industry in India. To appreciate the key priorities for such an effort, drawing on extensive interviews with representatives of the Indian electronics industry, the next chapter presents "The View From Industry."

The View From Industry— Regulations and Other Challenges

RESEARCH METHODOLOGY & INTERVIEW SAMPLE

This chapter presents findings of field research in India designed to shed light on the challenges faced, especially with regard to restrictive regulations, by India-based companies (both domestic and foreign) in the electronics manufacturing industry. The field research was conducted through semi-structured interviews with the following objectives:

- To explore the nature of work being done by electronics companies in India—including product mix, levels of value addition, technological complexity of products and processes, employment effects, and sourcing of technology;

- To understand how companies assess the challenges they face and how they define their strategic objectives;

- To examine how companies self-evaluate their management and technological capabilities—especially in electronics design and manufacturing;

- To identify regulatory barriers most constraining investment in electronics manufacturing and limiting the growth of India's electronics industry.

Between April and August 2013 a total of forty-six interviews were conducted in six cities. These interviews included representatives of thirty-nine companies, three government departments, three industry associations, and one non-governmental organization (NGO). Interviews were conducted with senior managers across a broad sample of India-based companies involved in various stages (electronic components, electronics manufacturing services, and final products; see Figure 4 below) of India's electronics manufacturing value chain.

By ownership nationality, the sample was fairly evenly distributed between Indian-owned companies (56 percent) and foreign-owned companies. The small number of start-ups interviewed reflects their limited presence, especially in electronics manufacturing. Finally, in geographic terms, almost two-thirds of the interviews were conducted in Bangalore, with an additional one-fifth in the Delhi National Capital Region.

Figure 4.

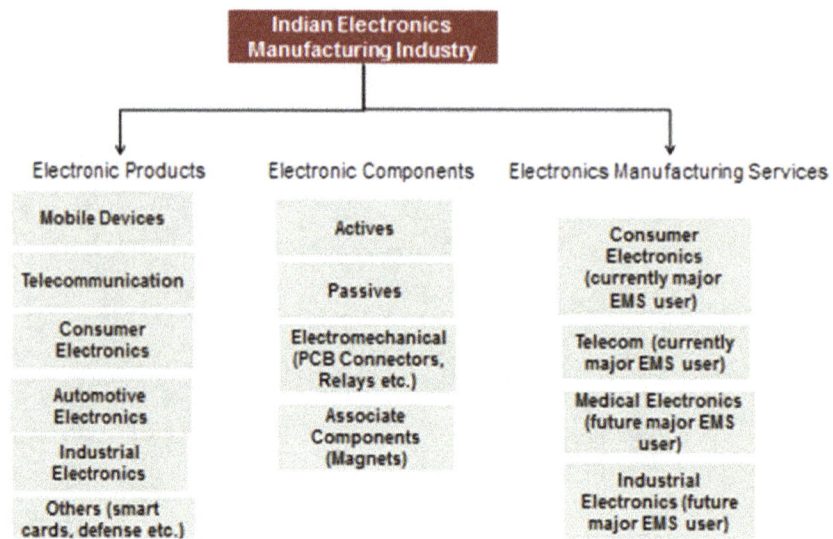

Source: Author's interviews

COMPANIES' STRATEGIES: CHALLENGES, OBJECTIVES, & CAPABILITIES

How Companies Define Their Strategic Objectives

Companies were asked to rank key motivations for investing in domestic electronics manufacturing as well as both enabling factors and challenges considered in deciding whether or not to do so.

Motivations. Start-ups and established companies differed markedly in identifying factors driving them to invest. For start-ups the primary reason was generally a response to customer needs—the ability to produce appropriate products to fulfil those needs. For the more established companies, besides serving customers, a major reason cited was frequently a need to increase market share.

Growing domestic consumption of electronic products was highlighted, throughout the interviews, as the main *enabling factor* which might convince companies to invest in electronics manufacturing. Underlying this is a desire to increase sales volume and reap economies of scale.[162]

Some of the more-successful companies emphasized a need to develop partnerships with global industry leaders as an important motivation for expanding investment in electronics manufacturing. Developing partnerships with global MNCs increases opportunities for exports and offers greater competitiveness in export markets.

Rangsons Electronics, a domestic EMS, noted:

For us the key reason for the expansion of manufacturing is the need to develop partnerships with global industry leaders. This has always been a major driver for us. For example, we are part of a global EMS alliance which gives us significant bargaining

Growing domestic consumption of electronic products is the main enabling factor which might convince companies to invest in electronics manufacturing

power in terms of getting preferred pricing in components. Similarly, we have global customer partnerships with companies like GE [General Electric] which means that Rangsons does the production for GE's global level products . . . and not just its products for the Indian market.

Skanray, a young domestic producer of medical equipment, emphasizes the importance of global partnerships as a source of technology and as a facilitator for gaining access to international markets:

A young domestic
producer of medical
equipment emphasizes
global partnerships as a
source of technology and
as a facilitator for gaining
access to international
markets

> For us the major reason for expanding manufacturing is to develop partnerships with global industry leaders. . . . Of course, gaining market share and responding to the customer's needs is also very important. Instead of talking about growing domestic consumption, I would rather talk about growing global consumption—which is an enabling factor for us.

Government policies and tax incentives were frequently mentioned but almost all interviewees complained about the absence of effective support policies and tax incentives. Even where companies knew about such policies there was widespread scepticism regarding whether they would be implemented effectively and lead to tangible results. Here are three typical comments:

- "The tax 'incentives' and government 'support' policies are all dissuaders rather than enablers"—Skanray Technologies (Indian start-up producing medical equipment).

- "Things like tax incentives and government-support policies have only been recently introduced and their effect remains to be seen. However, as a businessman, I only trust what I have seen and I would not rely on these incentives to be an enabling factor as long as they do not show tangible results"—Rangsons Electronics (Indian EMS).

- "The government-support policies and tax incentives are non-existent, so I would not call them enablers."—Ace Components (Indian component producer).

In terms of business strategy, most of the interviewees seek to compete as low-cost producers and predominantly for lower-end market segments—though a few companies aspire to follow closely behind global leaders with fast scaling-up. A handful of internationally oriented companies seek to work closely with a larges OEMs and produce items they require. Most companies, though not all, rate the level of local-value added as low.

How Strong are India-Based IC Design Capabilities?

As noted in the opening chapter, India's base of integrated-circuit (IC) design capability is one of its key potential advantages in the electronics industry. To understand the reality and impact of this capability, companies were asked how analog- and digital-design capabilities of India-based IC design companies compare with global best practices and to assess the level of sophistication of projects created in India.

This question provoked lively responses with most companies emphasizing a high level of technical-design capabilities but acknowledging a gap in the level of integration of design projects and the sophistication of project management. Overall the interviews confirmed the fundamental disconnect, highlighted in the opening chapter, between domestic electronics manufacturing and India's treasure trove of engineers with advanced electronic- and IC-design capabilities.

According to Cadence, a leading global electronic design automation (EDA) tool provider with a massive presence in India, the technical capabilities of senior and experienced Indian designers are on par with global leaders. MNC affiliates such as Intel and Texas Instruments conduct integrated design projects in Bangalore and such projects are estimated to account for around 70 percent of current projects against 30 percent for more basic design services. According to the same source, a serious concern is a quite significant gap in capabilities between MNCs and public sector units (PSUs). As compensation packages in MNCs are up to ten times higher than those in PSUs, MNCs hire the best talent.[163] Performance requirements and competition for jobs are much less intense in PSUs than in MNCs.

A more skeptical assessment was offered by Softjin Technologies, a domestic provider of EDA tools. According to this source, Indian IC-design companies are still peripheral players and require much catching-up to reach global best practices. This source has interviewed engineers who are working at MNCs and concluded that their "quality is not really good. International companies operating out of India are working on a manpower-supply model rather than a capability model."

In this view, MNCs still seem to emphasize primarily access to large populations of lower-cost young design engineers. MNCs have little interest in enabling Indian engineers to "own the full delivery of the chip. . . . The attitude is to look for manpower trading rather than getting into a solution mode for a particular problem."

A similar view was provided by the head of an affiliate of IBIDEN, a Japanese company producing electronic substrates: "The project sophistication handled is of medium level. Take the example of a company such as Intel. The design work they are doing in India is only the peripheral-level design. All the advanced core-design work is done in Israel."[164]

A common theme was to emphasize two critical weaknesses of India-based IC design teams: a) a "service mentality" focusing on detailed engineering implementation of designs received from global customers and b) a lack of exposure of Indian design engineers to strategic marketing and other business functions needed to develop and market original IC designs. Some companies also emphasized the weak design and R&D capabilities of SMEs.

REGULATORY BARRIERS: TAXES, TRADE, & LICENSES

Definition and Brief Overview of Regulatory Barriers

As noted above, this study builds on the World Bank's earlier analysis of India's restrictive regulations.[165] Companies were questioned at some length regarding regulations so as to identify which regulations are most constraining to electronics companies and their impact on the growth of this strategic industry.

MNCs emphasize primarily access to large populations of lower-cost young design engineers and have little interest in enabling Indian engineers to "own the full delivery of the chip"

To better understand relative priorities and impact, the concept of regulations was broadened from the earlier study to include such elements as trade laws and regulations affecting infrastructure provision (see Figure 5 below). Companies were asked which regulations they consider most constraining regarding investment in electronics manufacturing as well as related services and R&D and how these regulations related to other constraints.

Figure 5. Barriers to Growth: Taxonomy of Regulations Constraining the Growth of Electronics Manufacturing Industry

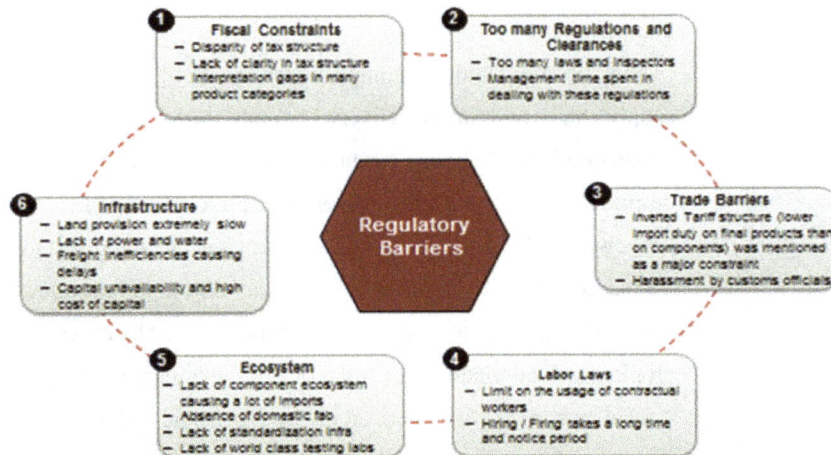

Source: Author's interviews

SMEs consider the high cost of capital as the single biggest constraint to investment. The negative impact of India's inverted tariff structure[166] also creates serious problems for both smaller and larger companies and was among the most-frequently cited barriers to investment in electronics manufacturing. The tariff structure has been treated in detail above and the cost of capital is considered beyond the scope of this study.

Fiscal constraints impact SMEs more than larger companies. Larger companies have better processes and can afford to use tax consultants to navigate through the complex maze of tax requirements. Frequently mentioned concerns relate to the instability of taxation, a lack of clarity of the tax structure, and the disparity of tax structure from state to state. This topic will be addressed in greater detail below.

Badly designed regulatory breaks and tax incentives are an important concern for SMEs—especially for young companies eager to get new ideas produced and introduced to the market. One example is perceived incentives to promising companies to stay small. According to the general manager of a Texas Instruments (a leading MNC) affiliate, "There are significant disincentives against scaling-up of SMEs. As soon as companies seek to get bigger, they are under pressure not to cross a ceiling, and entrepreneurs look for way to navigate these barriers, for instance by establishing shell companies."

Labor issues seem to be more of a problem for large companies—as micro, small, and medium enterprises (MSMEs) are typically working with a smaller workforce and a higher percentage of permanent employees.

'There are significant disincentives against scaling-up of SMEs. As soon as companies seek to get bigger, they are under pressure not to cross a ceiling'

Domestic-Market Fragmentation, Taxes, and the GST

There was a near-complete consensus among interviewees about the fundamental need for a nationwide GST. One interviewee compared waiting for GST with Samuel Beckett's *Waiting for Godot,* a play where two actors wait endlessly and in vain for the long-promised arrival of a man they know only by hearsay.

Companies both large and small mentioned the disparity between state taxes as a key fiscal constraint. This is by far the largest regulatory issue facing India-based electronics companies—the one mentioned first or second in almost all interviews. It was considered far more important than subsidies.

As addressed in the opening chapter, electronics manufacturing depends on complex global supply chains and timely delivery of goods. As long as a unified India-wide GST does not exist it seems highly unlikely that a robust electronics manufacturing industry can develop in India despite however many subsidies or other policies are generated for the industry. Most companies acknowledged the priority the government has already given to addressing GST concerns but they emphasized that the implications of further delays in introducing an India-wide GST will be felt across all segments of electronics manufacturing. The absence of an actual solution for this problem will continue to constrain one of the high-priority industries for India's future.

This was nicely summarized by the Electronic Industries Association of India (ELCINA) in its Budget Recommendations of November 2012:

> There has been considerable delay in implementation of GST regime in the country. As per promises made in recent months and expectation of industry, ELCINA strongly supports implementation of GST and hopes that the next deadline of 1st April, 2013 will be met. Implementation of GST, couched in simple language with maximum clarity will lead to all-round rationalization . . . ELCINA strongly supports immediate implementation of GST with CST [Central Sales Tax] subsumed in it. In case for some reason, the implementation of GST is delayed further, CST should be made zero.[167]

Beyond GST, the general difficulties in obtaining clarity on tax structure and regulations affect all types of companies equally hard, if in somewhat different fashions. Design start-ups (especially those companies lacking semiconductor wafer fabrication facilities) are strongly affected by the lack of clarity regarding their status. Manufacturing start-ups are hard hit by complex and costly tax-recording requirements.

An Inverted Tariff Structure as the Core Challenge

In almost all interviews, companies of all sizes, both domestic and foreign, confirmed an important finding from the second chapter of this study. They emphasized the negative impact of the inverted tariff structure—where duty on finished products is lower than that on high-tariff raw materials and intermediate products.

According to the Manufacturers' Association for Information Technology (MAIT), most materials used in electronics products (e.g., plastics, copper, aluminum, and ceramics) as well as components (both semiconductors and passive components) receive a customs duty of 10.3 per cent and a special additional duty of 4.4 percent.[168] The rationale for these duties is that such materials are of "dual use" since they may go into multiple products.

As long as an India-wide GST does not exist it seems highly unlikely that a robust electronics manufacturing industry can develop

The negative effects of the combined 14.7 percent duty are felt by all Indian electronic equipment producers. One, TVS Electronics, noted: "Currently all the components that are used in our manufacturing have a very high duty structure due to which the prices of our final products become uncompetitive. Import content on our products is close to 52 percent, and due to the higher duty structure on components, the overall disability is 30–40 percent on the final price of the product."

Interviewees, citing protection of national interest, in general supported India's decision not to join the July 2013 negotiations in Geneva (referred to as ITA-2) on broadening the scope and product coverage of the ITA.

Of particular interest is the negative impact of the inverted tariff structure on medical electronics, a "low-volume, high-value" industry segment.[169] Medical equipment requires complex technology and its production requires imports of electronic components and modules. This industry sector involves high levels of testing to meet statutory and regulatory requirements and requires highly skilled and experienced engineers and operators capable of running very costly high-precision machines.

Imports of medical equipment face a duty of 5 percent while materials for their production face a tariff of 5–7.5 percent—often with substantial delays caused by differing interpretations of whether an obscure regulation (Central Excise Annexure III, IGCRDMEG Rule 1996) is applicable. This inverted tariff would further increase for domestic producers if medical equipment was included under ITA-2 and India joined the agreement.

As reported by Skanray, an Indian start-up producing medical equipment, "The inverted tariff structure[,] which implies that the final products have no duty, is a major constraint for companies like us. Imports of Chinese medical equipment at ridiculously low prices have seriously impacted the growth of electronics industry in this country. . . . Hence, there is no level playing field for the different countries. For example, when Brazilian medical equipment is sold in India, they face a 7.5 percent duty while when Indian equipment is sold there it must face a 70 percent duty."

Customs Clearance

Even beyond the tariff structure, many interviewees complained about the operational details of the customs-clearance process which frequently delays production and turn-around cycles. Customs-clearance delays and their unpredictability differ across specific product segments but are significantly constraining in each. They may have a devastating effect in a fast-moving industry such as electronics manufacturing—especially for smaller and younger companies.

Companies interviewed reported that inefficient and corrupt customs procedures often delay customs clearance by at least two-to-three weeks. Given the ambiguous regulations, companies often face situations where customs officials seek to exploit this ambiguity to exact "informal payments." Companies have to struggle with continuing delays and with highly complex customs procedures for imports of supplies.

Interviews with two industry associations (Consumer Electronics and Appliances Manufacturers Association [CEAMA] and ELCINA) highlighted how these delays cause serious and systemic disruptions of the electronics industry's supply chain.[170] According to both, a root cause of these delays is that ground-level officers have discretion to decide on "notification-based exemptions" (i.e., customs provisions based on notifications which are time consuming to interpret and/or challenge).

A frequently cited example was the "Customs Notification 25/99," also known as the "jumbo notification."[171] This notification is supposed to list the raw materials going into electronics products and identify which of these raw materials qualify for receiving preferential duties. Interpreting this notification, however, causes seemingly endless delays and never-ending queries because of conflicting interpretations of the ambiguities of product classifications.

Suspicion has become built into the customs-duty system as well as opportunities for rent-seeking through "informal payments." Association leaders spend much of their time trouble-shooting consignments stuck at airports.

Customs officers often behave as if they believe all companies are trying to game the system. ICs, as an example, have a 0 percent duty while other parts face a 7.5 percent duty. Some companies may falsely declare components as ICs to reduce their overall weighted duty. Other companies may claim to import components as raw materials for electronics and then resell them. Clearly, companies engaging in such activities are in violation of the regulations and should be punished. Most companies and associations interviewed supported increasing punishments for violators.

The burden of proof is, arguably unfairly, heavily weighted against companies—which must fully satisfy all requests before being allowed to operate or access to their supplies. In principle, there is a "green channel" intended to allow companies with proven reputations to request and receive accelerated customs clearances—but this is only available to large companies.

The associations report that senior customs officers are responsive but are caught in a slow-moving system requiring far too many approvals. Especially for smaller companies, a grievance-redress mechanism capable of *fast* response is needed. The current administrative mechanisms are exceedingly slow and companies are scared of reprisals should they complain. As one interviewee put it: "To gain one rupee in customs duties the country is losing thousands."

As an example: The duty on LCD panels was cut from 10 percent to 0 percent. Some companies or countries, however, have an established practice of labeling "panels" as "modules." So when a shipment of LCD panels labeled as "modules" arrived, customs officers queried whether the duty should be 0 percent or 10 percent and refused to release the shipment without duty until they received a formal approval. Domestic production of a particular product line stopped as a result.

A CEAMA representative, going through DEITy, had to request that the Ministry of Finance issue a clarification notice to customs. The relevant joint secretary agreed but, before issuing the notice, that joint secretary was reposted. A new joint secretary arrived but was unfamiliar with the issue. The CEAMA secretary general had to make repeated visits, over a period of six months, only to be told each time "the file has moved," to clarify this simple issue.

Similar examples of dealing with customs authorities and interpretation of product names abound. There are conflicting interpretations of regulations concerning "metallic" versus "plastic"

The relevant joint secretary agreed but, before issuing the notice, was reposted. A new joint secretary arrived but was unfamiliar with the issue

materials with widely different implications for tariff rates. The point is not that these queries are not valid but regards the enormous complexity required to resolve them.

Reforming this system requires trust. One possible solution would be to allow consignments through but require bank guarantees on threat of forfeit. Larger companies do this but it is necessary to spread this system, as well, to smaller companies with established track records.

More generally, there may be cause for creating a new system to resolve tariff queries capable of providing answers in days rather than weeks. Such a system would be of particular benefit to electronics, which could be the pilot industry for such a mechanism which, when proven, would be spread to other industries.

A WEAK & DYSFUNCTIONAL STANDARDS SYSTEM

Many companies emphasized the critical importance of establishing quality and safety standards for developing India's domestic electronics manufacturing industry. Companies are conscious that effective standards are critical for market expansion and differentiation and are needed to facilitate technology transfer.

Most companies emphasized that well-defined quality and safety standards could be a powerful policy tool against low-cost and low-quality imports. There was also a broad consensus that India's current laws and standards-development structure are insufficient to guarantee the high levels of quality and safety essential for the industry's international competitiveness.

In line with this broad concept of standardization as a tool for industrial development, some companies argued that India should study China's approach and develop a unified national standardization strategy.[172] Other interviewees referred to the voluntary standards system approach used by the United States as a possible benchmark, noting its emphasis on public-private partnerships in standards development.[173]

To compete as preferred suppliers within global production networks, India-based companies need well-designed interoperability standards enabling "two or more networks, systems, devices, applications, or components to exchange and readily use meaningful, actionable information—securely, effectively, and with little or no inconvenience to the user."[174]

Given the importance of this subject, the following chapter will examine in detail the broader strategic role standards can play in fostering the growth of India's electronics industry.

THE CHALLENGES OF YOUNG INNOVATIVE COMPANIES

A critical finding from the interviews was that smaller companies face a disproportionate burden from being exposed to the maze of restrictive regulations. This is especially true for young companies seeking to produce new products but struggling to cope with existing tax, customs, and myriads of other regulations.

A major challenge for any company considering investing in electronics manufacturing in India is that the company needs to immediately comply with a multitude of clearances needed for establishing a manufacturing facility. Smaller companies, especially, are overwhelmed by the multiplicity of regulations and the time needed to cope with these requirements.

India's current laws and standards-development structure are insufficient to guarantee the high levels of quality and safety essential for the industry's international competitiveness

Time required for having land allotted and obtaining clearances often adds between six and eight months in India. In China similar clearances would require no more than two-to-four weeks.

As of this writing, the Consortium of Electronic Industries of Karnataka is working on a fifty-acre cluster where the land allotment has not been cleared for more than eighteen months. During this time the government in Karnataka changed and the entire process had to be restarted.

A large number of interviewed companies mentioned that, even though they seem to have slightly improved, excise and commercial departments continue to remain a major clearance hurdle. Companies providing services for the electronics manufacturing industry complain about the service-tax department and the excessive time it takes to receive credit for any possible overpayment of service taxes.

Over and above these examples, the following concerns are familiar from many media accounts and other studies and were frequently mentioned: lack of coherence, excess paperwork, numerous and often unclear laws, and the interaction of corruption with all of these.

Promising start-up companies able to raise early-stage venture capital are hard pressed to make sense of well-intentioned but complex incentives and support policies provided by the Ministry of MSMEs and other government agencies. Many of these companies lack the scale and the deep pockets needed to cope with the substantial compliance costs of existing regulations and the multifaceted and often obscure tax and tariff obligations. Nor can these companies afford the production delays generated by inefficient customs-clearance and transportation systems.

Existing regulations, bank lending, and support policies fail to address the needs of companies seeking to draw on their product-development strengths and system integration to pursue "low-volume, high-value" strategies.

These findings are consistent with those of one of the few empirical studies on India's technology-based start-up companies.[175] Drawing on interviews with a sample of 443 start-up companies, that study found that "government policies represent the greatest problem faced by start-ups in India."[176]

Among these policies, the same study (drawing on a separate slightly larger sample of 532 companies) prioritized the specific constraints: lengthy procedures, formalities, and extensive paperwork (20 percent); high import and excise duties and sales tax (20 percent); stringent norms of labor laws (15 percent); interpretation of laws and policies by enforcement agencies (14 percent); stringent environmental- and pollution-control norms (12 percent); various insurance plans (10 percent); frequent raids and inspections (8 percent); and taxation (2 percent).[177]

In addition to the constraints identified in the prior research, interviews for this study highlight the paucity of institutionalized support for technology-based start-up companies in the electronics manufacturing industry. Despite progress in de-licensing and de-regulation, India's framework conditions for innovative start-up companies remain weak.

IMPACT ON BUSINESS ORGANIZATION

An important finding from this study is that persistent restrictive regulations may give rise to forms of business organization which prevent organic growth through the accumulation of specialized resources and capabilities. Deeply entrenched restrictive regulations clearly play a major role in constraining growth and stifling innovation. Some companies interviewed establish "shell companies" for the sole purpose of by-passing the effects of tax, labor, and other regulations or for availing themselves of subsidies and other plans.

All the companies which described such activities did so with the condition of confidentiality. The common features, though, of such activities are easily described. One legal company serves as the hub of the web. This company will typically have fewer than ten employees—principally legal and accounting experts. It will buy and sell from a web of shell companies, maintaining low sales income and profits on its books but accumulating residual assets. The shell companies conduct the actual trading, each shell processing few enough sales to keep their exemption from excise taxes and employing few enough people to remain under the labor-law thresholds.

Every few years the shell will be rotated out of circulation, its sales slowing, to stay under the radar. If a new plan offers subsidized land to small companies, new shell companies will be established to take advantage such benefits. If inspectors query why six companies operate out of a single address and undertake the same activities then the companies reach an "informal arrangement" ensuring the questions proceed no further.

As one proprietor described it, managing these underground webs is time consuming but, under current regulations, it can be very lucrative. When a business is sold, only the hub company is sold, the shells becoming, in effect, defunct unless the new owner decides to maintain the existing shells rather than creating a new web.

These shell companies are very different from the complex network arrangements established by Korean and, especially, Taiwanese electronics companies. Those network arrangements legitimately function to generate economies of scale and scope in procurement, marketing, manufacturing, and R&D.[178]

In India's electronics industry, avoiding regulations is the sole purpose of these sometimes-quite-complex network arrangements. The result is that much of the management effort within these "shell companies" concentrates solely on keeping alive the increasingly complex "underground" web of companies actually making money from short-term contracts. Such networks, however, lack the human and financial resources to invest in plants and equipment—much less R&D. These networks fall far short of the minimum economies of scale and scope required for competing in the fast-moving and technology-intensive electronics industry.

Much of the management effort within these 'shell companies' concentrates solely on keeping alive the increasingly complex 'underground' web of companies actually making money

Support Policies

KEY QUESTIONS & FRAMEWORK FOR RECOMMENDATIONS

The final chapter of this study examines the Indian government's support policies seeking to fast-track the growth of India's electronics industry. This study develops recommendations to strengthen the implementation of the government policies and to address observed gaps. Given the previous findings on policy parameters and the business environment for electronics manufacturers, the following questions have guided research for this chapter:

- How familiar are India-based electronics manufacturers with the government's policies in support of their industry?

- What needs to be done to improve the impact and the effectiveness of these support policies?

- What additional policies are required to fully utilize strategic standards in fostering the growth of electronics manufacturing in India?

- How do India-based electronics manufacturers evaluate the recently launched National Policy on Electronics?

- How do India-based electronics manufacturers rate the effectiveness of existing support policies in terms of their creating an enabling environment for electronics industry growth in India?

Key information sources include responses and information received through interviews with company and association representatives described in the prior chapter, descriptions of key elements of the NPE described in detail in policy documents published by a variety of government agencies, reports prepared by industry associations and consulting companies, and prior academic papers on specific aspects of India's support policies. These are complemented by information from the author's earlier research on electronics-industry support policies in China, Korea, Taiwan, and the United States.

After presenting findings for the above questions, this report concludes by offering recommendations distinguishing between "priority actions" and "fundamental process changes."

"Priority actions" are actionable specific changes in regulation or support policies, e.g., implementing a national GST, staying out of ITA-2, increasing awareness of NPE, and cutting tariffs on components.

"Fundamental process changes" involve ongoing longer-term changes, e.g., revamping India's standardization system and the structure of standards-development organizations, generating more sophisticated trade diplomacy, implementing a different inspection and dispute-resolution regime for customs, and improving the existing architecture for electronics-industry working groups.

A distinction is also made between "outward facing" and "domestic" recommendations.

"Outward facing" recommendations relate primarily to trade and FDI policy as well as to interactions with global organizations, such as the WTO, and with international industry associations, such as private standards consortia and the World Semiconductor Council.

"Domestic" recommendations relate primarily to regulatory changes, support policies, and the interaction between government and the diverse segments of India's electronics industry.

Such distinctions are not binary opposites as some actions include aspects of multiple categories. Discussion of "standards" provides an example: standards are both outward facing, being tools of trade policy and globally set, and inward facing, as means to induce higher quality and improved safety locally. Standards also require discrete actions, with specific standards urgently needed in such areas as medical electronics, as well as requiring ongoing process, as the structure of standards-development organizations needs to be revamped and attitudes to standards shifted to become more strategic.

These categories are also interdependent: most priority actions will not have a long-term effect without process changes but, similarly, process changes will be difficult to implement without priority actions creating momentum for change. Tariffs and trade policy will have little long-term impact if companies must continue to spend much of their management focus arguing with customs if an LCD module is or is not duty free—or managing a complex web of shell companies solely to circumvent regulations.

INDUSTRY PERCEPTIONS, IMPLEMENTATION, & DIALOGUE

Interviewed companies were asked what indicators they would identify to assess whether a support policy might facilitate an expansion or upgrading of their manufacturing business. Criteria selected differed by industry segment as well as by the size and ownership pattern of the companies.

By itself, this observation should convey an important message to policy makers—the diversity of interests of different stakeholders in electronics manufacturing requires a diverse set of policies and a capacity to listen to the voices of different industry stakeholders.

At the same time, certain priority concerns were widely shared. Almost all respondents emphasized the critical importance of transparent and user-friendly support policies formulated with industry input.

The remainder of this section presents industry's responses and then, in line with the Twelfth Five-Year Plan, discusses the general constraints of policy implementation; highlights restructuring industrial dialogue as a key mechanism for industry input and feedback and policy guidance (with a deep and detailed comparison to the government/industry dialogue in Taiwan); and leads to the role of industry associations—key enablers for such dialogue.

Companies' Specific Expectations for Support Policies

Most interviewees strongly emphasized the need to attract and facilitate investment by both domestic and foreign companies to increase the local-value-added component of their businesses. Many companies emphasized that investment promotion should heavily focus on attracting investment from large foreign MNCs as such investment acts as a catalyst in creating space for smaller domestic companies. A number of companies argued that particular attention should be given to attracting EMS providers.

The two most-mentioned expectations (see Figure 6 below) involved reducing general infrastructure constraints and improving access to capital. The third-most-mentioned expectation, and the highest priority specifically relevant to the electronics industry, is that industrial-support policies should accelerate the entry of start-ups and foster the rapid growth of young India-based companies seeking to manufacture and commercialize new electronic-hardware products. Successful policies would enable domestic companies, especially SMEs, to develop their own intellectual property rights.

Figure 6. What Industry Expects from Policies in Support of Electronics Manufacturing and Innovation

Top wish list

Listed by decreasing number of mentions	Provide the basic manufacturing infrastructure: Land, power, water, etc.
	Reduce the cost of capital and make capital easily available
	Foster innovation by funding innovative start-ups
	Reduce the multiplicity of regulations through single-window clearance
	Invest in skill development and knowledge transfer
	Push the fab policy
	Invest in standardization and testing facilities
	Foster and build a domestic component industry

Source: Author's interviews

The Indian Electronics and Semiconductor Association (IESA) provided concrete suggestions for industrial upgrading policies.[179] The association suggests strengthening the technological and management capabilities of companies through support policies encouraging international technology cooperation with such foreign centers of excellence as Germany's Fraunhofer Institute and Belgium's IMEC International[180] and the facilitation of technology-licensing agreements.

IESA also suggested that industrial-support policies should induce high value-added FDI to address the rapid domestic market growth (mobile phones as an important growth market); foster co-creation, where cooperation between companies (both MNCs and domestic companies) and India-based academic institutions will lead to co-owned IPR . . . . A fundamental prerequisite for implementing this strategy is "the development of sophisticated IPR protection, IPR licensing and IPR development capabilities."

Interviewed companies were also asked which states (or cities or clusters) already have the infrastructure, human resources, capabilities, and supportive policies strongly positioning them for growing investment in electronics-hardware manufacturing. Figure 7 (below) summarizes the results of their responses. The leading three states on this list might come as little surprise but some low-income states ranked unexpectedly high on the list, most notably West Bengal. Companies also placed emphasis on the importance of looking at clusters, as much or more than at states, in determining the potential for growth.

Figure 7. Which States Have a High Electronics Manufacturing Potential?

Which states already have the underlying factor endowments (infrastructure, human resources, capabilities) that could in theory position them well, but require substantial reforms and/or investments to unlock this potential (e.g., low-income states which could show strong growth in the medium- to long-term)?

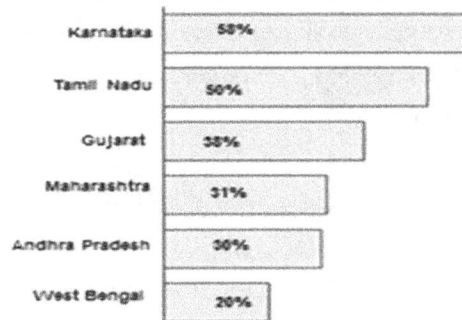

State	Percentage
Karnataka	58%
Tamil Nadu	50%
Gujarat	38%
Maharashtra	31%
Andhra Pradesh	30%
West Bengal	20%

Source: Author's interviews

The General Constraint of Transparent, User-Friendly Implementation

Almost all respondents placed much emphasis on the ease of policy implementation and the transparency and "user-friendliness" of such policies. Many companies expressed frustration concerning support policies which may be brilliant on paper but are obstructed by weak implementation by both central and state governments.

Related was the concern that support policies be predictable and offer a longer-term perspective based on a thorough analysis of competitive dynamics and the relentless and unpredictable pace of technical change. Most companies agreed that one-time initiatives to fix regulatory constraints may be useful but also lead to a dispersal of effort and risk rapid obsolescence as the industry changes so quickly. Given the relentless and unpredictable pace of technical change, in three years' time this year's reform may no long have a significant effect or may even be harmful.

A fundamental challenge for regulatory reform is a capacity for *flexible* policy implementation which, based on a periodic review of what works and what doesn't, can be recalibrated and existing regulations adjusted.[181]

This response is directly aligned with a core theme of the Twelfth Five-Year Plan—that the issue in India is not the intent of policies but their implementation. As the plan states:

> Two root causes for poor implementation are: inadequate consensus amongst stakeholders for policy changes, and very poor coordination amongst agencies in execution . . .
>
> . . . [Often], . . . a good plan was not made before announcing action. Or, a plan was made but it was not understood by, and sometimes not even known to, the various

The issue in India is not the intent of policies but their implementation

agencies involved. And, often, even when the plan was known, there was no monitoring and follow-up.

The Planning Commission specifically highlights a range of *implementation constraints* for a cohesive national manufacturing strategy. These range from the complexity of inter-ministerial and state/central government relations to the multiplicity of stakeholder groups which must be involved for a plan to have traction.

Fundamental changes are required in the management of government programs to overcome these deeply entrenched implementation constraints. According to the Planning Commission, the government should shift from a role of micro-manager to one with capabilities focused "not only on scheme design and strategic alignment of schemes to tactical outcomes, but also strong evaluation and feedback systems and networks from which the states and other local implementers can learn."

Support policies for industrial manufacturing need to focus on learning and capability development: "A good manufacturing plan focuses on accelerating learning within a country's industrial ecosystem that enables enterprises within it to improve their competitiveness faster than enterprises in other countries. The implementation system for such a plan needs to focus on building broad-based capabilities across industries."[182]

Most fundamentally, the Planning Commission suggests improving inter-agency collaboration, establishing effective stakeholder consultation processes, and continuous evaluation of policy impacts and effectiveness. The Planning Commission calls for "wide-spread consensus-building processes . . . [which] . . . must become part of the Indian manufacturing system. For this, institutions for representation, such as employee unions, employer associations, and civil society organizations, must become more professional, more democratic, and more competent in arriving at agreements that ensure fairness to all stakeholders."[183]

These extracts highlight the critical importance of changes in the processes of policy implementation. Making policies relevant to industry needs requires permanent "industrial dialogues" on many levels and more direct access to entrepreneurs, a willingness of government agencies to listen to industry needs (companies complain that they need to "chase" the authorities), and encouraging the development of mission-oriented public-private partnerships.

According to many respondents, upgrading India's electronics manufacturing industry will require multiple industrial dialogues, focusing on specific projects and outputs, between industry and government.

The call for such dialogues is not new. Such recommendations can easily become nothing more than clichés extolling the virtue of "public-private dialogue." To make this recommendation more concrete it will be useful to refer to one of the more outstanding examples of such dialogues, those which have taken place in Taiwan.

Industry Dialogues and Policy Innovators: The Example of Taiwan

Taiwan has generated a multi-layered system of industrial dialogues. Its achievements in the electronics-system design-and-manufacturing industry would be impressive for any economy—they are even more impressive for a small island, about one-third the size of New York state in

Upgrading India's electronics manufacturing industry will require multiple industrial dialogues, focusing on specific projects and outputs, between industry and government

the United States. With a population of roughly eighteen million people in 1980, less than half that of South Korea's thirty-eight million for the same year,[184] Taiwan lacked a large and sophisticated market, lacked specialized capabilities and support industries, and lacked the science and technology infrastructure necessary for manufacturing and developing technologically demanding electronics products. Initially SMEs, with limited resources and capabilities and limited capacity to influence pricing or shape the development of markets and technological change, dominated the Taiwanese electronics industry.

To overcome the dual disadvantages of working from a small economy and consisting largely of SMEs, Taiwan's electronics industry's policies quickly developed strong links between government-supported research institutes, industry associations, and private industry. The Industrial Technology Research Institute's (ITRI's) Electronics Research and Service Organization (ERSO) was critical in fostering technology co-development, its diffusion and use for commercial-scale manufacturing, and the creation of multiple domestic and international industrial dialogues.[185]

Industry associations played a vital role as initiators, enablers, and coordinators of industrial dialogues. Of particular interest to India is the *international* orientation of Taiwan's industry associations in the electronics industry. Consider the Taiwan Semiconductor Industry Association (TSIA), which seeks, among other goals, not only "to promote cooperation among different sectors in the local semiconductor industry" but also "to participate in global standard setting and activities related to the development of the semiconductor industry." It identifies its key role as serving a bridge, as much or more than lobbying, seeking to "engage in international negotiations on behalf of the local industry" and also to "create better communications among its member companies and with other industry associations."[186]

For another example consider the Chinese American Semiconductor Professional Association (CASPA).[187] Founded in 1991 as a professional association of Taiwanese semiconductor engineers, CASPA has developed into the largest Chinese American semiconductor professional organization. Worldwide, CASPA consists of more than four thousand individual members, corporate sponsors, a board of directors, a board of advisors, a board of volunteers, and honorary advisors. Headquartered in Silicon Valley, CASPA has nine local chapters worldwide and more than seventy corporate sponsors, from EDA and design companies to foundries, venture capital, science-and-technology parks, and legal and financial-service companies located in China, Hong Kong, Japan, Singapore, Taiwan, and the United States.

CASPA's international orientation has made an important contribution to the exposure of Taiwan's electronics industry to leading-edge technology and management practices. It has also provided an excellent mechanism for worldwide networking and knowledge sharing.

The promotion of associations of this nature, as well as their integration into the policy process, has immensely strengthened the ability of policymaking in Taiwan to respond realistically to global trends.

In recent years it has become clear that Taiwan's earlier industrial-development model is reaching its limitations. Appropriate new policies are now needed to develop new domestic Taiwanese capabilities for low-cost innovation at both company and industry levels.[188] In

In Taiwan, industry associations played a vital role as initiators, enablers, and coordinators of industrial dialogues

response, Taiwan has devised a new electronics industry policy combining market-led innovation and public-policy coordination of multiple layers of industry through dialogues between private-industry and public stakeholders.

Given its pragmatism and openness to new forms of public policy and private-public partnerships, Taiwan's new policy may, in fact, serve as a benchmark for India's attempts to enhance the impact of India's support policies for the Indian electronics industry.

Taiwan's Ministry of Economic Affairs (MOEA) promotes individuals to become "policy innovators." These are government officials who are incentivized to not only design a particular regulation but to take personal responsibility in enabling, across the life cycle of the regulation, its effective implementation and ensuring flexible adjustments to the initial regulation are made where necessary.

The ministry's Industrial Technology Development Program generates various mission-oriented working groups for specific projects such as "smart electronics" or "streamlined manufacturing technology." These meet regularly in specialized forums with mid-level participants from academia, associations, government agencies, and industry developing and implementing pre-competitive cooperative-research agendas and implementation schedules.[189] These groups seek to integrate the R&D resources of academia, research institutes, and industry.

Another feature of Taiwan's industrial-policy process in the electronics industry includes the outsourcing of many policy-making functions to the Taiwan Institute of Economic Research (TIER) and the Chunghua Institution of Economic Research (CIER). These are think tanks created by MOEA tasked to design and implement policy and regulations in cooperation with industry and industry associations.

Noteworthy features of the Taiwanese system are the regular and repeated use of "committees" for consensus building among ministries and experts from academia, industry associations, and private companies as well as the creation of "seminars" on highly advanced topics which are actively used for content-based interaction with the private sector.[190]

Taiwan's policy process is characterized by a proliferation of institutions coming together in frequent, repeated, and multi-level forums oriented towards programs, projects, and substantive content with a globally-facing, international, orientation throughout. This has created Taiwan's capacity, based on thorough preparation, for gradual evolution of industrial policy and for flexible adjustments responding to changes in markets and technology.[191]

This model differs from India's ad hoc and personality-driven "advisory councils," composed of eminent individuals, where personal dynamics have frequently dictated policy formulation. For instance, the "technical evaluation committees" for NPE's Modified Special Incentive Package Scheme (M-SIPS) contain mostly government officials, one or two academics (often from the same institution), and a few representatives of large and established companies. Missing are India's diverse electronics-industry associations as well as representatives from smaller and start-up companies.

It is important, however, to emphasize recent positive examples indicating moves towards broader industry representation in NPE policy initiatives.

Noteworthy features of the Taiwanese system are the regular and repeated use of "committees" for consensus building among ministries and experts from academia, industry associations, and private companies

The "Brainstorming Session on Indigenous Product Design and Development of Digital Set Top Boxes" (May 9, 2012) was organized by the Office of the Principal Scientific Advisor to the Government of India and the Indian Electronics and Semiconductor Association (IESA).[192] Out of the forty-three members of this working group, twenty-nine are from industry with representation from three industry associations (CEAMA, Confederation of Indian Industry [CII], and IESA).

Another example was the "Meeting to Ascertain the Manufacturing Capabilities of LED/LED based Lighting Products in India"[193] convened by DEITy (May 17, 2013). Out of seventeen participants, twelve were from industry with representation from the above-noted three industry associations.

It is critical that such initial steps are strengthened and furthered, becoming regular and repeated dialogues focused on tangible actions as opposed to a series of one-time efforts.

The Role of Industry Associations

As noted above, strong industry associations are a *sine qua non* of a robust industrial dialogue. Reflecting this, many interview respondents argued that the role of industry associations deserves much greater attention in formulating policy and regulations. Within India's electronics industry there is a surprisingly large number of over-lapping associations at both the national and state levels giving rise to significant fragmentation. Too many associations with over-lapping constituencies and mandates reduces the voice and influence of each individual association.

These associations differ substantially in the services they provide and in their implementation capacity. Associations such as ELCINA and CEAMA devote a large share of their limited resources to fixing problems caused by "ground-level implementation of regulations and corruption" and to navigating import consignments through custom authorities. In addition to their focus on fixing the endless delays, queries, and quarrels of the import process these associations lobby for subsidies to counter what they estimate to be a 10–12 percent "disability cost" on domestic electronics manufacturing.

ELCINA, however, has also been active in shaping and implementing NPE's electronics manufacturing cluster (EMC) policies. ELCINA is working with three clusters: outside Delhi, in Bangalore, and in Chennai.[194] ELCINA is seeking to overcome the lack of inter-agency communication within the Indian government such as the problem that present EMC grants cannot cover housing since a) public housing is a separate mandate of a different ministry and b) there is fear of subsidized housing speculation. In general ELCINA seeks to provide industrial-dialogue services ranging from disseminating information to training programs and seminars.[195]

A similar role is sought by the Indian Electronics and Semiconductor Association (IESA). Since 2009 IESA has been intentionally seeking to co-shape the NPE. It has broadened its membership to include both leading MNCs and domestic companies active across the semiconductor value chain. ISEA argues that their membership and efforts have had a positive effect on improving the sophistication and pragmatism of policy concepts in the NPE (including the fab policy, on which there is more below).

IESA focuses on three primary efforts: events, industry research, and government interface. It highlights the following "Industrial Dialogue" objectives:[196]

The role of industry associations deserves much greater attention in formulating policy and regulations

- Create global awareness of the Indian semiconductor and electronic-systems industries beyond the generic "IT" umbrella
- Create win-win interactions among semiconductor and electronics-product and services companies, government, academia, venture capitalists, and industry bodies
- Create an enabling ecosystem catalyzing industry growth and leadership
- Foster active collaboration between industry and universities to further expand the available world-class semiconductor talent pool.

ISEA seeks to reduce possible trade conflicts with major trading powers by disseminating information on key initiatives of India's NPE (seeking, for instance, to play a mediating and transmitting role in the controversial PMA).

It remains to be seen whether ISEA is claiming greater influence than it has in reality or whether ISEA can play a useful role in recalibrating government policies to the requirements of the WTO and the need to attract technology transfer through FDI.

Both ELCINA and ISEA are considered simply as examples. Others—including MAIT, CEAMA, and the Consortium of Electronic Industries of Karnataka—have also been discussed earlier.

A complex and difficult question is what types of policies or other actions support the development of capabilities in such associations so that they may become strong partners in developing public policy. One element must be the type of government response: if dialogue results more often in subsidy plans rather than difficult structural reforms such associations will simply privilege lobbying for subsidies. Many companies currently report that industry associations don't play a role and some companies have tense relationships with association managements.

As the implementation of the NPE proceeds and as more difficult topics are addressed it may be useful to consciously shape association responses, improve responsive internal governance within the associations, and involve them in more difficult tasks. One of the most difficult tasks to be addressed will be creating and implementing a more strategic role for industry standards.

THE STRATEGIC ROLE OF STANDARDS

Why Standards are Critical for Latecomer Industrialization

Standardization is often perceived as primarily a technical issue receiving only limited high-level policy support. Technical standards, however, contribute at least as much as patents to economic growth.

As a key mechanism for the diffusion of technological knowledge, technical standards contribute to productivity. The macroeconomic benefits of standardization thus exceed the benefits to companies alone.

A widely quoted study conducted for the German Institute for Standardization (DIN) found that a 1 percent increase in the stock of standards is positively associated with a 0.7–0.8 percent change in economic growth.[197]

Such econometric studies only scratch the surface. Equally important are qualitative impacts on environment, food, health, and work safety. These broad qualitative impacts of standards are essential for successful economic development—a well-functioning standardization system and strategy serves as a catalyst for translating new ideas, inventions, and discoveries into productivity-enhancing innovation. Standards are a critical link in growth strategies seeking to create quality jobs in higher-value-added advanced manufacturing and services.[198]

There are, however, a potentially infinite number of standards—each differing in form and purpose. This poses a demanding challenge for countries newly beginning to develop standards systems and strategies. Rapid and disruptive technical change (such as the transition to the "Internet of Everything"[199]) creates new challenges for standardization.

Of critical importance are interoperability standards needed to transfer and render useful data and other information across geographically dispersed systems, organizations, applications, or components.[200] This process has increased the economic importance of standardization especially for emerging economies such as India. India remains a latecomer to industrial manufacturing and innovation but is, at the same time, deeply integrated into international trade, capital markets, and foreign direct investment.

An operational definition addressing standardization within the electronics industry will be useful. One state-of-the-art definition may be taken from the US National Institute of Standards and Technology (NIST; developed as part of its Smart Grid Interoperability Standards project).[201] Standards are:

> [s]pecifications that establish the fitness of a product for a particular use or that define the function and performance of a device or system. Standards are key facilitators of compatibility and interoperability. . . . Interoperability . . . [is] . . . the capability of two or more networks, systems, devices, applications, or components to exchange and readily use . . . meaningful, actionable information—securely, effectively, and with little or no inconvenience to the user. . . . [Specifically, standards] define specifications for languages, communication protocols, data formats, linkages within and across systems, interfaces between software applications and between hardware devices, and much more. Standards must be robust so that they can be extended to accommodate future applications and technologies.

At the most fundamental level, standards help to ensure the quality and safety of products, services, and production processes and to prevent negative impacts on health and the environment. Standards enable companies to reap the growth and productivity benefits of increasing specialization.

Today, however, specialization extends well beyond trade into manufacturing and services including engineering, product development, and research. Equally important is the international dimension. As globalization extends beyond markets for goods and finance into markets for technology and knowledge workers, standards are no longer restricted to national boundaries. Standards become a critical enabler of international trade and investment—they facilitate data exchange as well as knowledge sharing among geographically dispersed participants within global corporate networks of production and innovation.[202]

Standards are the lifeblood of latecomer industrialization. For countries such as India a robust system of developing technical standards is necessary not only to reap economies of scale and scope but also to reduce transaction costs and to prevent duplication of effort.

Standardization has become a complex and multi-layered activity involving multiple stakeholders differing in their objectives, strategies, resources, and capabilities. Standardization is a highly knowledge-intensive activity requiring well-educated and experienced engineers and other professionals. While engineers originally created this discipline, key concepts are now shaped by legal counselors as well as corporate executives and government officials.

This implies that an effective system of standardization for India's electronics industry requires close cooperation between industry, government, academia, and non-governmental organizations representing the broader interests of society. Even within industry, different stakeholders with conflicting interests reflect differences in size, ownership patterns, and business models as well as whether companies are standards users, implementers, or developers.

Latecomer Standardization is Costly

As well as sophisticated processes and a variety of skills, considerable financial resources are required to develop and implement effective standards. A rough estimate of such costs may be gained from a stylized model distinguishing important tasks of standardization and highlighting differences in capability sets and standardization strategies.[203] Table 3 highlights important tasks of standards development. Typically tasks 1, 3, and 4 are the most costly. However, in cases involving litigation, legal costs can easily run into the hundreds of millions of dollars in the United States.

Table 3: A Taxonomy of Standardization Tasks[204]

1. Develop technology to support the standard
2. Analyze the cost-benefit ratio of adopting an existing international standard versus creating a new domestic standard
3. Identify licensing fees for essential patents (for both existing standards and for newly created standards)
4. Pass testing, conformity assessments, and certifications
5. Identify membership fees for formal and informal standards-development organizations
6. Quantify logistics (travel, etc.) costs
7. Assess the cost/risk of including one's own patents within a general standard
8. Manage the patent pool
9. Establish back-end support for standards implementation and quantify costs
10. Establish legal (litigation) support and quantify costs
11. Establish lobbying support and quantify costs

As for the capabilities required to undertake these tasks, consider a simple model distinguishing between two countries.

Country A (the "innovator") has a long history of standardization, a proven ability to operate successfully within standardization bodies and to shape international standards, a fairly diversified production and innovation system, and a broad base of accumulated knowledge and intellectual property rights (IPR) helping to generate product and process innovations. Country A thus is able to "control much of the technological input necessary to meet the standards."[205] As a result, a primary concern of law and policies in country A is the protection of IPR and the "openness" of standards is subordinated to IPR protection.

Country B (the "latecomer") lags behind in the development of standards. Country B is a standard taker, manufacturing products that are developed and standardized by country A. Country B still has to learn how to operate successfully within standardization bodies. Most importantly, country B still has a long way to go to establish a fairly diversified production and innovation system and a broad base of accumulated knowledge and IPR to allow it to shape, or at least co-shape, international standards. As a result, a primary concern of law and policies in country B is to focus on economic development and the diffusion of the knowledge inherent in IPR. Standardization is viewed as an enabling platform for innovation and latecomer economic development.

In principle, these countries and their companies may select one or a combination of the standardization strategies described in terms of rising sophistication in Table 4. Country A and its leading companies would likely pursue "standards leader" or "co-shaper" strategies while country B and its leading companies would likely focus on "free rider" or "fast follower" standardization strategies.

Table 4. International Standardization Strategies

Free rider: Let MNCs decide which standards to use, and save cost by not investing in the development of domestic standardization infrastructure and capabilities

Fast follower: Develop domestic capabilities that are necessary to rapidly adopt existing standards so that the standard's technology can be manufactured and marketed quickly

Co-shaper: Revise existing standards and/or adapt proposed new standards to suit domestic needs so that current and new products may be quickly developed and marketed

Standards leader: Create new standards to suit domestic needs and embed essential domestic patents within these standards so that current and new products may be quickly developed and marketed

Latecomer economies, such as India, face opportunities and challenges in their standards and innovation policies differing considerably from the opportunities and challenges faced by today's advanced economies. Latecomers typically are standards takers and have far to go to begin to shape or even co-shape international standards. Latecomers are also typically more vulnerable to

Latecomer economies face different challenges in their standards and innovation policies than today's advanced economies

"strategic patenting" strategies permitting patent holders to require usage fees through their control of de facto industry standards.

Latecomer economies also lag behind advanced economies in their standardization capabilities and are likely to face higher incremental costs in developing and disseminating effective standards. Ubiquitous globalization and rapid and disruptive technical change (such as the increasing complexity of digital networks) create new challenges for standardization.

No Indian electronics company can succeed in international trade without mastering the interoperability standards necessary to transfer and render useful data and other information across geographically dispersed systems, organizations, applications, or components. The response to this challenge is not a one-time, pseudo-optimal, "national strategy for standards" or similar effort. What is needed is the development of a strategic approach towards standards development allowing and incorporating opportunities for continuing adjustments.

In a world of increasing complexity and uncertainty, it is always preferable to have built-in redundancy and freedom to choose among alternatives rather than seeking to impose from the top the "one best way" of doing things. Increasing complexity drastically reduces the time available for standards development and implementation. This makes it practically impossible to get solutions right the first time. There may have to be many policy iterations, based on trial and error, and an extended dialogue with all stakeholders to find out what works and what doesn't.

Increasing complexity also makes it more difficult to predict all possible outcomes of any particular policy measure—especially unexpected negative side effects (of which there may be an endless variety). A small change in one policy variable describing a particular procedure for achieving compliance with a particular standard may have far-reaching and unexpected disruptive effects on other policy variables and outcomes.

It is difficult-to-impossible to predict the full consequence of standardization policy interactions among an increasingly diverse population of both domestic and international stakeholders. Given the diversity of competing stakeholders in standardization issues, the result of a particular national standards policy will depend more on negotiations and compromises than on logical clarity and technical elegance.

India's Standardization System

How well does India address the requirements for standardization of latecomer industrialization? While standards systems in China, Korea, and Taiwan have been extensively studied[206] there has been, to the author's knowledge, no systematic study of India's national standardization system.

A recent background study prepared for the US National Academies was intended to assess the status and challenges faced by India's standardization system. Unfortunately that study provides no more than a superficial description of existing institutions without analyzing their systemic challenges and weaknesses.[207]

The Bureau of Indian Standards (BIS) is India's official national standardization and certification body.[208] BIS oversees the development of Indian Standards (IS)—coordinating, through its technical committees, input from various public and private-sector stakeholders. Today there are over eighteen thousand standards in the Indian market.[209] Beyond Indian national standards there are many other types of standards in use in India including those developed by the

In a world of increasing complexity and uncertainty, it is always preferable to have built-in redundancy and freedom to choose among alternatives rather than seeking to impose from the top the 'one best way' of doing things

International Organization for Standardization (ISO), the International Electrotechnical Commission (IEC), and other international standards developers as well as regional standards, foreign national standards, and more.

One of India's major challenges is to overcome the complex and highly fragmented institutional organization of India's Standards and Conformity Assessment Bodies. There are a number of standards-development organizations (SDOs) overlapping significantly in terms of objectives, responsibility, and authority.

The Department of Science and Technology (DST) promotes new areas of science and technology and related standards and functions as a nodal department for organizing, coordinating, and promoting science and technology activities within the country.[210]

The Quality Council of India (QCI) is designed as an autonomous body by the Indian government to establish and operate a national accreditation structure for standards-conformity assessment bodies.[211] The National Accreditation Board for Testing and Calibration Laboratories (NABL)[212] provides accreditation services for testing/calibrating laboratories in accordance with ISO/IEC 17025.[213] The National Accreditation Board for Certification Bodies (NABCB)[214] undertakes assessment and accreditation of certification bodies applying for accreditation as per the board's criteria consistent with international standards and guidelines. The National Quality Control (NQC) organization is then also responsible for spreading awareness on advantages of compliance to quality standards and continuous improvement.[215]

A number of standards institutions compete for resources and responsibilities within the electronics industry. These include the Electronics and Information Technology Division Council (EITDC)[216] of the Bureau of Indian Standards (BIS), the Telecommunications Engineering Center (TEC),[217] the Global ICT Standardization Forum for India (GISFI),[218] and the Development Organization of Standards for Telecommunications in India (DOSTI).[219]

Of these latter standards-development organizations, GISFI and DOSTI seem to be most active, involving participation both from industry and academia. It is noteworthy, however, that in interviews for this study most of the respondents were not or only vaguely familiar—with the exception of BIS, TEC, and DOSTI—with most of these organizations.

Thus it is hardly surprising that the recently inaugurated "Seconded European Standardization Expert in India," representing Europe's leading standardization organizations,[220] stated that "India's standardization system remains very complicated and the Technical Regulations are still very intertwined with Technical Specifications, so it makes it sometimes quite difficult for European exporters to understand the requirements that apply to their products."[221]

This confusion regarding the intent of India's standards system and its division of labor was shared by most of those interviewed. They reported that India's existing standards system is weak—with the BIS considered under-resourced and not interacting adequately with industry. Overall the respondents felt strongly that the existing standards system needed to be strengthened and upgraded, both financially and organizationally, so that companies may receive effective support for the use of quality standards and certification requirements.

One of India's major challenges is to overcome the complex and highly fragmented institutional organization of India's Standards and Conformity Assessment Bodies

Government Initiatives

Until recently, the critically important field of standardization has remained a "white space" in policy development.

India's government acknowledges that bold changes are necessary in the organization of the country's standardization system. So far, however, the primary focus of policy-making has been on developing robust safety standards and certification rules which function as trade policy instruments. There is still a fundamental neglect regarding the strategic role technical standards may play as a tool for developing and upgrading India's electronics manufacturing industry.

A report prepared for the Twelfth Five-Year Plan states unequivocally: "Lack of domestic regulations and standards is a potential cause of import of substandard goods that may not only put our consumers and environment at risk but also leads to an unfair and cut-throat competition for the domestic manufacturing industry."[222]

As a signatory to the WTO-Technical Barriers to Trade (TBT) Agreement, India agreed not to implement technical requirements creating unnecessary obstacles to international trade. However, as the report states, the "Agreement provides flexibility for member countries to specify the requirements in the interest of national security, prevention of deceptive practices, protection of human health or safety, animal or plant life or health, or the environment."

Taking advantage of this flexibility, though with a somewhat narrow concept of India's standardization strategy, the above report recommends the following policy initiatives:

Mandating Standards. BIS must, in a phased manner, be strengthened to ensure availability of Indian standards for every finished electronic good and to mandate compliance with safety, electromagnetic compatibility, and Restriction of Hazardous Substances (RoHS) standards. DEITy needs to be strengthened to regulate standards and ensure compliance with established standards. DEITy should "create a specialized enforcement wing to handle fake, spurious, [and] non-complying goods."

Upgrading Standards Infrastructure. A robust infrastructure is necessary for mandating, testing, and certifying standards by establishing recognized test and certification laboratories through public-private partnerships and in cooperation with international agencies such as ISO, IEC, and other international standards consortia.

Implementation Budget. To implement the reform of India's standard system and to ensure effective enforcement of standards, the report suggested a budget of Rs 275 crore (approximately US$40 mn[223]) over the Twelfth Five-Year Plan to ensure compliance of electronic products to standards of safety and electromagnetic compatibility. Specifically, this budget would finance the establishment and upgrading of test and certification laboratories and support companies in covering the cost of accreditation, mutual-recognition plans, and participation in international standardization meetings and technical committees. This budget would also be expected to cover the cost of standards education-and-training programs for industry participants, customs, and border-control agencies.

On the basis of these recommendations, DEITy issued a number of notifications and administrative guidelines defining requirements for the compulsory registration of standards, the procedures for setting-up and upgrading electronic product-testing and certification laboratories, and a quite confusing number of additional administrative regulations.[224]

These standards-related regulations culminate in the government's "Electronics and Information Technology Goods (Requirement for Compulsory Registration) Order, 2012," which came into force on July 3, 2013. The primary strategic objectives of the policy are:

- Providing Indian consumers with the right to enjoy world-class goods.

- Upgrading the quality of domestic products to achieve global competitiveness.

- Developing strategies to combat the dumping of non-compliant goods.

- Projecting internationally a positive image of India as a country producing high-quality electronics and IT goods.

Recommended Additional Actions

This review indicates that, although there has been a substantial increase in India's attention to standards, more can and should be done given the importance of standards and the still-unmet needs. Suggestions for immediate-priority actions include:

First, substantially invest in standard-setting for priority products such as medical devices. This will improve the competitiveness of Indian companies against low-cost, low-quality, imports—especially those from China. Such standards should be set high, both to ensure safety and to create a source of discipline for domestic companies needing to meet such stringent quality levels to benefit from the protection from low-quality imports.

Second, address the opportunities for cost savings in standards development, testing, and certification. Lacking established standards, companies must repeatedly test for every customer and all regulatory purposes to demonstrate quality.

Third, identify companies which are actively developing, implementing, and using national and international quality standards and provide such companies privileged treatment regarding access to incentives and eligibility for government support.

Adopting the level of standardization capabilities as the discriminating criterion should support the effectiveness of such government policies and make it possible to move beyond industry-wide incentives to considering the values of individual cases. High standardization capabilities could be used to select companies which should be primary recipients of incentives and industrial support.

Fourth, look beyond the Indian national borders to learn from already-developed best practices in policies, procedures, and organizational approaches. Standards-development organizations in the European Union, Japan, Korea, Taiwan, and the United States are all eager to deepen their links with India's standards-development organizations—especially in the information technology and electronics industries.

Consider creating an "India—European Institute of Standards and Innovation" with two campuses: one in India (possibly linked to one of the Indian Institutes of Management [IIM] or the Indian Institute of Science [IISC] in Bangalore) and one in Europe (possibly linked to the European Telecommunications Standards Institute [ETSI]).

Although India's attention to standards has increased, more can and should be done given the importance of standards and the still-unmet needs

Such an institute would train engineers, executives, technicians, government officials, and academics from both India and the European Union. The institute could provide technical consulting services enabling both Indian and European companies to solve problems arising from differing standards systems. Similar forms of international cooperation should be considered with the Institute of Electrical and Electronics Engineers Standards Association (IEEE-SA) and other private standards consortia eager to strengthen their position in the Indian electronics industry.

As senior Indian government officials have acknowledged, India faces a capability gap in standards-development that it does not face in other areas of electronics-policy formulation. In establishing semiconductor policy, for example, there are countless talented Indian engineers—whether local or diaspora—providing substantial input. Indian engineers have been less active and less visible in international standards-setting bodies. India must build capabilities for a strategic approach to standards and begin establishing the processes and dialogues to harness and support those capabilities.

NATIONAL POLICY ON ELECTRONICS (NPE)

Objectives and Policy Tools

India's National Policy on Electronics was established within the context of India's National Manufacturing Plan.[225] The National Manufacturing Plan has five objectives including both the creation of one hundred million jobs and increasing technological depth.[226] The planning commission has called for a "new policy paradigm" to implement the manufacturing plan. The primary role of the central government is intended to be to provide opportunities for multi-layered "industrial dialogues" to cope with the increasingly complex coordination of networked industrial manufacturing.

> Thus, the paradigm of policy planning in manufacturing must shift from "planning as allocations" to "planning as learning"; and from budgets and controls towards improving processes for consultation and coordination. In India we have already given up the paradigm of allocations and quotas and there is no question of reverting to it. However, having not mastered the other paradigm yet, we are not able to grow our manufacturing sector as fast as we could.[227]

A related policy, the National Telecom Policy (NTP),[228] was approved in May 2012 and includes initiatives such as free roaming, national number portability, a unified licensing regime for operators, and a push for expanded broadband usage.

Of critical importance for India's electronics manufacturing industry is that the NTP seeks to increase domestic manufacturing of telecom equipment. It seeks to "promote indigenous R&D, innovation and manufacturing to reduce dependency on imports and enhance exports." The specific goal is a "complete value chain for domestic production of telecommunication equipment to meet Indian telecom sector demand to the extent of 60 percent and 80 percent with a minimum value addition of 45 percent and 65 percent by the year 2017 and 2020 respectively."

NTP seeks to use preferential procurement, "consistent with key trade agreements," as well as stronger standards and IPRs to achieve these ends.[229] While these are all laudable and long-overdue initiatives, NTP does not provide specifics regarding when and how such initiatives will be implemented.

The Indian National Policy on Electronics was approved in October 2012 and seeks to create "a globally competitive Electronic System Design and Manufacturing (ESDM) industry." The stated objective was to increase "domestic production" to US$122 bn by 2017, with US$20 bn in exports.[230] Expected direct employment was projected to be three-and-a-half million workers while indirect employment effects were expected to add up to six-and-a-half million workers.

As published in November 2012, NPE identified significantly higher objectives for the ESDM industry without specifying how much would actually be attributable to electronics manufacturing. The ESDM industry is targeted to achieve, by 2020, a turnover of US$400 bn involving investment of about US$100 bn, ESDM exports of US$80 bn (up from US$5.5 bn in 2012) and employment of around twenty-eight million workers.[231]

In addition to those quantitative targets, the NPE seeks to accomplish a range of qualitative goals including: improving governance mechanisms, creating robust institutional mechanisms for standards, fostering frugal innovation, and supporting innovative start-ups.

These are ambitious objectives when contrasted with the development of electronics manufacturing industries in China, Korea, and Taiwan. India seeks to build an integrated domestic electronics value chain in less than ten years—a process that took decades in China, Korea, and Taiwan.

To achieve these objectives the NPE sets forth the following eight policy priorities:

1. Provide incentives for investment through a Modified Special Incentive Package Scheme (M-SIPS)

2. Establish semiconductor wafer fabrication facilities

3. Provide preferential market access (PMA) to domestically manufactured electronic products

4. Provide incentives for establishing two hundred electronics manufacturing clusters (EMCs)—establishing greenfield EMCs and upgrading brownfield EMCs

5. Establish a stable tax regime and market India as an attractive investment location

6. Create a completely secure domestic cyber ecosystem

7. Implement e-waste (management and handling) rules

8. Establish a national electronic mission.

The proposed NPE budget for 2012–2017 is Rs 33,375 crore—or almost US$5 bn. Almost a third is allocated to the semiconductor wafer initiative, almost two thirds are allocated to various incentive and infrastructure plans, and the remainder is allocated to other initiatives.

Industry Perceptions

An important finding is that most companies know very little about the specific policy tools of NPE and their current implementation. A typical response was that the announcement of NPE signals a long overdue step in the right direction. But most interviewees reserved judgment on the

An important finding is that most companies know very little about the specific policy tools of NPE and their current implementation policy itself as they lacked information on when and how those policies would be implemented.

Interviewees were most aware of PMA and of the Electronics Development Fund (EDF) and rated such plans as potentially useful, provided they were properly executed, for developing the industry. Only a few interviewees raised concerns regarding non-compliance with WTO obligations nor were there suggestions the government needed to develop new and sophisticated policy approaches to trade, FDI, international standards, and trade rules. This would appear to reflect the limited international orientation of India's electronics manufacturing industry.

The wafer-fabrication policy attracted much attention but most interviewees confessed that they knew little regarding its current implementation. The fab policy generated varying responses as to how effective wafer fabs would be in developing the industry and what specific types of fabs deserved priority.[232]

Noteworthy exceptions to the lack of NPE awareness were a few companies having a material interest in the development of a domestic electronics manufacturing industry. This may be illustrated by the following response from the representative of Synopsys, an international EDA tool provider:

> Yes, I am familiar. In fact, I was part of the report that was published by DEITy in 2009 (the task force report which was a joint effort by IESA, ELCINA, MAIT, etc.). We had given ninety-nine recommendations, which were cut down to fourteen by DEITy. Five out of these fourteen recommendations can be seen in the NPE which are: EMC (electronics manufacturing cluster), fab policy, Modified—Special Incentive Package Scheme (M-SIPS), PMA, and EDF.

It was widely acknowledged by interviewees that, at least on paper, a holistic approach is now in place. There are, however, concerns that the government is primarily driven by macroeconomic factors and the widening trade gap and that it neglects the fundamental structural flaws of India's electronics manufacturing industry—primarily its weak and incomplete ecosystem.

According to the representative of SLN Technologies, a domestic EMS provider, "[t]he Indian electronics hardware industry has a lot of missing links which are retarding its growth—and only setting up of the whole ecosystem can stop that. Ideally this should have been done twenty-five years ago but better late than never. Look at the textile industry; we have the whole ecosystem and thus the industry is generating employment and is successful. The same is the case with automotive industry; it is doing most of the value addition in India itself. But the work done by the electronics industry is restricted to low-value-adding activities like assembly."

There was a widespread consensus that components manufacturing is the most critical industry bottleneck and the one which should receive priority attention from the NPE. Most interviewees, however, emphasized a gradual approach to building a more integrated electronics ecosystem.

There was a widespread consensus that components manufacturing is the most critical industry bottleneck and the one which should receive priority attention from the NPE

Priority Products

The NPE provides two somewhat conflicting objectives for the government's choice of priority products. One is the legacy of India's public-sector defense-electronics complex. A stated objective of NPE is "[t]o progressively increase the domestic production of the requirements of strategic sectors, namely, defense, atomic energy and space through domestic production, through appropriate combination of public sector and private sector."[233]

A second NPE objective emphasizes "large-volume" production, identifying these five priorities: cheap "budget smartphones," set-top boxes (STB), flat-panel displays (FPD) and tablets, optoelectronics (and especially LED), and smart meters (for deployment in smart electrical grids).[234]

The underlying assumption is that India's huge demand for these devices and the projected rapid growth of their domestic markets will induce massive investments in large-scale production lines generating economies of scale and scope and enable domestic production to compete with China-based mass-production lines.

As discussed in the first chapter of this study, there may still be some space in a few select industry segments for late entrants, such as India, to focus on "high-volume, low-cost" production lines. However, global transformations in information technology and markets are defining a new manufacturing imperative for India. Unlike China and other earlier industrial latecomers from Asia, India can no longer rely *exclusively* on "high-volume, low-cost" manufacturing as its main strategic option for expanding its electronics manufacturing industry. India's NPE should, instead, seek to create an alternative industrial manufacturing paradigm—"low-volume, high-value" production.

A focus on low-volume, high-complexity products would face less immediate competition from China and would leverage India's strengths in IC design and related capabilities

In the interviews only a minority of respondents clearly favored the "low-volume, high-value" manufacturing paradigm. These respondents believed a focus on low-volume, high-complexity products would face less immediate competition from China and would leverage India's strengths in IC design and related capabilities. Examples of implementing such "low-volume, high-value" manufacturing included frugal innovations in medical equipment and domestic production of strategic and defense products.

The overwhelming majority of interviewees, however, insisted on emphasizing large-volume production, as "high volumes will incentivize [the] development of [the electronics] component ecosystem," as "components account for a huge chunk of India's import bill."

Some respondents offered a more differentiated view—including specific suggestions. Some, both domestic smart-card producers and foreign (Japanese) producers of substrates, suggested adding smart cards to the NPE list of priority products. These respondents, however, also noted that, at present, almost half of the domestic capacity for smart cards is lying idle. They felt policies needed to focus on developing security standards providing disciplined protection for such production, rather than attempting a "race to the bottom" requiring government subsidies for cost-competitive production.

Another strong suggestion was to add medical devices to the list. Supporting this suggestion is the rapid growth of the Indian medical-devices market. This market was valued at Rs 17,742 crore (US$3.3 bn) in 2011[235]—but imports currently cover nearly three-quarters of this demand. India faces a huge and rapidly growing need for affordable medical devices to cope with its most pressing health needs. India's strengths in IC design could facilitate entry into the design and development of India-specific medical devices meeting the specific requirements of local conditions, i.e., "frugal innovation."

The Semiconductor Wafer Fab Policy

India's attempts to establish domestic wafer fab lines have a long but checkered history. In 1983 the government decided to establish a full-scale wafer fab, called the Semiconductor Complex Limited (SCL), in Chandigarh.[236] Despite high expectations the project failed to become a substantial aspect of the wafer fab industry. In the early 1980s SCL entered into a technical collaboration with American Microsystems Inc. but the fab remained far behind the leading-edge in wafer size and process technology.

In 2007 the government made yet another attempt and announced an ambitious plan to foster wafer fabrication along with the production of products such as photovoltaic (PV) solar cells and LCDs. The initial response was limited. While the government received proposals worth US$6.2 bn, none were for setting up chip manufacturing. In fact SemIndia, a consortium created to bid for a US$3 bn chip-making facility in Hyderabad, never submitted its proposal.[237] Negotiations with Intel, which had indicated an interest in investing in a wafer fab line in India, ended in failure. Intel claimed the "government dragged its heels on introducing an investment policy on semiconductors."[238]

In response, and seeking to learn from these prior attempts, the NPE semiconductor-wafer-fabrication policy was designed by a committee established, in April 2011, to identify technology and investors for setting up two semiconductor-wafer-fabrication manufacturing facilities.[239]

"Fab-1" was intended to use "established technology to support fabrication of varieties of chips to meet the requirement of high volume products as well as the requirement of the fab-less design companies on pay per use basis. This activity may involve either setting up a plant in India with established technology or acquiring an existing fab abroad and its relocation to India. The Government support needed for either of the options would have to be negotiated."[240]

In contrast, "Fab-2" was to be set up "as a green field cutting edge state-of-the-art facility. This would require provisions for giving equity/grant to an established Integrated Device Manufacturer to establish its fabrication facility in India. The amount of equity/grant would have to be negotiated."[241]

The committee estimated that the two fabs would require an investment of roughly Rs 25,000 crore (around US$5 bn), and added: "The exact level of Government support could be finalized by way of negotiations. The Government support could be by way of equity/grant/subsidy in physical/financial terms."[242]

As the wafer fab policy is a signature plank of the NPE, it received substantial attention during the interviews. Company representatives were specifically asked:

- Would it facilitate your business if local fabs existed, i.e., would it reduce your cost, time-to-market, and would it facilitate investment in new product development?

- What type of fab would be most conducive for your efforts to expand and upgrade your operations?

  o mature-process technology based on second-hand fab equipment

  o analog fab

  o leading-edge digital wafer fab with minimum investment costs of US$4 bn

- Is it realistic to focus on leading-edge fabs? Should India try to mobilize the huge investments required for such leading-edge technology given the extreme volatility especially of markets for memory devices?

- Would it be more realistic to focus on analog fabs needing less-advanced process technology than leading-edge digital wafer fabs?

Most interviewees accepted the strategic rationale for investing in a *diverse portfolio* of domestic fabs to reduce the existing unsustainably high import-dependence for semiconductors. Interviewees believed establishing wafer-fabrication lines in India would help to reduce the cost and the time needed to procure semiconductor chips when compared to securing them from abroad. This would be especially useful for manufacturing start-ups.

As semiconductors are strategic components, having virtually no domestic Indian manufacturing capabilities was expected to lead to continuing over-reliance on imports. The huge and growing import costs for semiconductors would be reduced if domestic wafer-fabrication capabilities were established. The establishment of wafer-fabrication facilities would be a critical step in creating an integrated value chain for electronics in India. Experience gained by Indian design companies working in close coordination with domestic fabs would add significantly to their design capabilities.

High capital expenditure and the excess capacity available globally would make it very difficult to generate a fast payback on investment in leading-edge domestic wafer fab lines

Many interviewees, however, also expressed concern whether an effective execution strategy is in place to cope with substantial implementation barriers. Not only would India need to import the extremely costly fab production equipment, it would also need to import the intangible knowledge needed to cost-effectively run the fabs. The overall life-cycle cost of running a leading-edge fab would likely be enormous—in an extremely cyclical and often quite unpredictable industry.

India's poor logistics network might make domestic procurement require more time than procuring chips from Taiwan or even China. Excess capacity already exists at most fabs in China and Japan so running competitive profitable fabs might not be easy.

There was widespread concern that the high capital expenditure involved and the excess capacity available globally would make it very difficult to generate a fast payback on investment in domestic wafer fab lines, especially those for leading-edge 450mm wafers with 22nm and below technology.

One risk for such a venture comes from power disruption—even slight disruptions in power supply can have devastating effects on yields and may require costly and time-consuming recalibration of equipment.

A serious challenge is the extremely high water consumption required by wafer fabs. A leading-edge wafer fab today "uses anywhere between 2 to 4 million gallons of very, very pure water—we call it ultrapure water—per day, and that, on the average, is roughly equivalent to the water usage of a city of maybe 40,000 to 50,000 people."[243] In light of India's severe water shortage,[244] this is hardly attractive.

Water shortages are especially severe in Bangalore,[245] initially one of the primary candidates for locating India's wafer fab. On August 26, 2013, the state government of Karnataka announced: "Much as we would have wanted the prestigious project to be based in Bangalore, which has perhaps the best ecosystem for electronics manufacturing in the country, we would be unable to host it because of the heavy demand it would place on water resources."[246]

Still another risk is that India, as a latecomer to wafer fabrication, will need both time and significant investments to develop a capacity for handling the toxic wastes produced by wafer fabrication.[247]

Some interviewees emphasized that establishing diverse domestic fabs will take time. All the more important would it be, they felt, to develop a portfolio of diverse policies with different time perspectives—with careful selection of pilot projects to produce rapid results.

At the time of writing, no final decisions have been made and decisions and policy statements continue to be in flux. On September 13, 2013, the Indian government, chaired by the prime minister, has given "in-principle approval" of two competing consortia offers of establishing two chip fabrication units.[248]

Few details have yet been announced. According to India's telecom minister, "Cabinet has in-principle also approved that incentives that will be given to these players, will be offered to other players (as well) who are interested in setting up semiconductor plant here . . . Incentives . . . are already covered under existing policies, which account for about 62 [percent] and the balance 38 [percent] is in form of loan provision, which is refundable. The burden on government will be only interest charges."

Formal announcements have yet to come. Government support for these units still must be negotiated with chip makers. This is as far as the 2007 negotiations with Intel proceeded before they went astray.

A decision that one of the fabs would produce analog devices would offer substantial advantages:

- Cost effectiveness: Analog fabs are much more cost effective than digital fabs. While the digital fab may cost billions of dollars in just the setup costs, not to mention the millions to be spent in operational expenses each year, an analog fab can be set up in the cost range of hundreds of millions of dollars.

- Close coordination and design skills: Analog chip design involves close coordination with the chip manufacturer so, in this regard, having a local fab can help tremendously in growing the capabilities of the Indian design industry.

Such a pragmatic approach is in line with research on the economics of wafer fabrication. Successful leading-edge 300mm wafer-fabrication facilities typically require US$9–12 bn in annual revenue.[249] Revenue requirements are substantially higher for the emerging 450mm wafer-fabrication facilities. India is at least five-to-ten years away from becoming such a market.

If one takes into account the widely discussed technical challenges facing even the global industry leader Intel in its transition to 450mm wafers with 22nm and 14 nm technology it becomes clear why leapfrogging to 22nm technology is an unrealistic goal for India when compared to a strategy of building capabilities through diversified fabs.

Industry Views on other NPE Support Policies

M-SIPS. The M-SIPS plan is, on paper, an ambitious incentive package to "offset disability and attract investments in large-scale manufacturing in the Electronics System Design and Manufacturing (ESDM) Industries.[250] The plan "provides subsidy for investments in capital expenditure—20 percent for investments in SEZs [special economic zones] and 25 percent in non-SEZs. It also provides for reimbursement of CVD [countervailing duty]/excise for capital equipment for the non-SEZ units. For high-technology and high-capital-investment units, like fabs, reimbursement of central taxes and duties is also provided. The incentives are available for investments made in a project within a period of ten years from the date of approval."

For the period until 2020, the NPE seeks to attract investments of around US$100 bn, leaving open, however, how much of this should go specifically into electronics manufacturing. For FY2013–2014 a target has been set to attract investments of Rs 25,000 crore (approximately US$4 bn). By August 2013, however, the government has received investment proposals worth only roughly Rs 4,600 crore (or US$700 mn) awaiting clearance, less than 20 percent of the budget allocation.

In contrast to the optimism projected by the ministry, most interviewees confessed to knowing little about the NPE's Modified Special Incentive Package Scheme (M-SPIPS) and, thus, were reluctant to share their assessments. As M-SPIPS is a central building block for implementing NPE, this low awareness is a worrying finding.

The Electronics Manufacturing Cluster (EMC) plan. One of NPE's important objectives is to "provide world-class infrastructure for attracting investments in the Electronics Systems Design and Manufacturing (ESDM) sector . . . , [to] . . . encourage development of entrepreneurial ecosystem, drive innovation and catalyze the growth of electronics manufacturing."[251]

The proposed electronics manufacturing clusters (EMC) plan "would support setting up both greenfield (new) and brownfield (existing . . .) EMCs."[252] For greenfield EMCs, "the assistance will be restricted to 50 percent of the project cost subject to a ceiling of Rs 50 crore (US$10 mn) for every 100 acres of land. For brownfield EMCs the assistance will be restricted to 75 percent of the project cost subject to a ceiling of Rs 50 crore."[253]

Company representatives were asked in their interviews whether the proposed EMC plan could replicate the earlier success of similar plans for software and how it should differ from them. Most respondents cited major problems with the EMC plan. They noted its potential for creating land scams, the unrealistic selection of cluster locations, and the unattractive conditions for EMCs—especially start-ups. A minority of respondents agreed that clusters are necessary as

Most respondents cited major problems with the EMC plan—noting its potential for land scams, unrealistic cluster locations, and unattractive conditions for EMCs—especially start-ups

islands of good infrastructure in a developing economy and pointed to the success of India's car-components industry as well as the huge benefits Chinese and Taiwanese companies have historically reaped from such manufacturing zones.

The EMC plan replicates many of the features of the Scheme for Integrated Textile Parks (SITP) originated by the ministry of textiles. This plan has been notably more successful than prior programs in developing industrial parks and zones in India and has, in fact, avoided some of the problems that respondents identified.[254] This could indicate that—in line with a common theme—electronics companies are not fully aware of the details of the EMC program. EMCs do, however, differ from SITPs in crucial aspects, including allowing a more active role in location selection by organizations not themselves entrepreneurs (e.g., state industrial development corporations and associations). These differences may be the cause for the reported unattractive locations of the currently projected EMCs. If this is, in fact, the case, it would justify tweaking of the EMC program—ideally through the type of industrial dialogue described above.

The Electronics Development Fund (EDF). This potentially important NPE policy tool could help strengthen the weak innovation capacity of India's electronics manufacturing industry. It would create a dedicated fund to support seed, angel, and venture funding with an initial Rs 5,000 crore (approximately US$800 mn). This proposed fund is now with Ministry of Finance and, at present, is still under discussion.

A draft proposal, published by DEITy in November 2012, highlights the following objectives:[255]

> There is an urgent need for intervention to promote and develop innovation, R&D, Indian IPR and manufacturing within the country for electronic products, which include telecom products, especially those having security implications. . . . The fund may be leveraged to acquire foreign companies so as to shift the production of products currently imported in large volumes, into the country. Some of the PSUs which are well positioned may take a lead role and venture into such acquisitions. The fund would be managed professionally and accessible to both Government and private sector.

Specifically, the EDF proposal recommends funding an extensive list of priority activities related to electronics R&D including:[256] the design and fabrication of an Indian microprocessor, the creation of a "manufacturing value-addition fund" to provide interest-linked subsidies linked to domestic value addition, a seed fund to support start-ups, a fund to provide multiplier grants for industry-academia linkages, a focused venture fund, and an equity/venture fund to nurture solar PV start-ups and research projects.

However, there are only a few suggestions on how administrative processes and communication with industry would need to change to facilitate an efficient and speedy implementation of the EDF proposals.

Company representatives were asked in their interviews what they know about the current status of EDF and how the EDF should be organized to best facilitate the entry of innovative start-up companies into India's electronics manufacturing industry.

Most interviewees acknowledged that they knew little concerning the details and current status of the EDF plan. Industry representatives voiced strong expectations mixed with substantial doubts regarding whether such an ambitious plan could be successfully and fairly implemented.

Such doubts should not be taken as a fundamental barrier. Foreign venture-capital funds are expressing interest in the EDF and it may yet become one of the highest-impact initiatives within the NPE.

What is clear is that, if a fundamentally strengthened implementation process is needed anywhere, it is with the EDF. Committees and working groups needed for its implementation deserve sustained consideration, tilting strongly towards the newer models of industry dialogue rather than the older "business as usual" models discussed above.

LAST THOUGHTS: THE GROWING IMPORTANCE OF INTERNATIONAL TRADE DIPLOMACY

The second chapter, on "Policy Parameters," documented the negative impact of India's inverted tariff structure on the growth of India's electronics manufacturing industry. The third chapter, "The View From Industry . . .," documented that almost all interview respondents singled out the inverted tariff structure as a major barrier to investment in this industry.

Clearly policies to upgrade India's electronics industry need to place considerable effort on developing smart approaches to international trade diplomacy. It must be emphasized that trade diplomacy has an important domestic component. The findings of this study indicate that it is time now for the government to reconsider whether the IT services sector still needs priority attention in trade-agreement negotiations. One could argue the focus now needs to shift to the domestic electronics manufacturing industry and its needs.

A new approach to trade diplomacy would focus on negotiating revisions in the Information Technology Agreement (ITA) acknowledging and correcting the asymmetric effects ITA has had on cost structures and capabilities in India's electronics manufacturing industry.[257] Plurilateral agreements such as ITA should allow for special and differential treatment of latecomers.[258]

Trade-related policy tools developed as part of India's NPE fall far short of such an agenda. In India's *Foreign Trade Policy Plan,* released May 2012 by the director general of foreign trade and additional secretary to the government of India, the electronics industry ranks only tenth out of fourteen industries.[259]

Priority trade-policy initiatives for the electronics industry may largely build on existing policy:[260]

a. Export of electronic goods shall be incentivized under the focus product scheme.

b. Expeditious clearance of approvals required from DGFT shall be ensured.

c. Exporters/associations shall be entitled to utilize the Market Access Initiative (MAI) and Market Development Assistance (MDA) government programs for promoting electronics and IT hardware manufacturing industries' exports.[261]

d. Electronics sector shall be included for benefits under the government's Status Holder Incentive Scrip (SHIS) scheme.

India's engagement with the institutions shaping the global economy is a critically important complement to high-profile efforts to build wafer fabrication facilities and provide government financial incentives

Regarding India's participation in ITA and in FTAs, the NPE does not seem to have introduced new initiatives addressing India's new trade-diplomacy requirements. Echoing the interview suggestions again, this fundamentally important area would seem to be a natural place to deploy new and strengthened mechanisms for industrial dialogue. Such a concerted effort could also reap substantial symmetries with an outward-looking, strategic approach to standardization. In combination, these efforts could provide disciplined and smart protection as well as push Indian companies and institutions into more active roles in international-standards bodies.

India's engagement with the institutions shaping the global electronics industry is a critically important complement to high-profile efforts to build wafer fabrication facilities and provide government financial incentives. Such a two-pronged strategy could provide India with an enduring and sustainable boost to its electronics manufacturing industry.

Notes

[1] Department of Electronics and IT (DEITy), *Electronics e-Newsletter,* 3, no. 24 (October 2013).

[2] Electronic Systems, Design & Manufacturing Ecosystem: Strategy for Growth in India (New Delhi: Ernst and Young, 2009), p. 9.

[3] Ibid., p. 10.

[4] Frost and Sullivan and IESA, 2013. See also http://forbesindia.com/article/briefing/indias-electronics-import-bill-could-become-larger-than-its-oil-bill/32386/1; http://www.business-standard.com/article/sme/increased-import-of-chinese-goods-hurts-indian-msmes-113061701112_1.html; http://www.deccanherald.com/content/319130/electronic-goods-import-up-30.html .

[5] Ministry of Commerce and Industry press release, *India's Foreign Trade: November 2013,* December 11, 2013, http://commerce.nic.in/tradestats/filedisplay.aspx?id=1 .

[6] See detailed analysis below.

[7] Hayao Nakahara (N.T. Information Ltd., Japan), presentation at India Printed Circuit Board Association, November 2008, http://www.ipcaindia.org/pdffiles/PCBMktOppDrNakahara.pdf .

[8] Ibid.

[9] http://coai.in/docs/Booz%20Study%20on%20Equipment%20Manufacturing%20Policy.pdf .

[10] For details on India's market structure in the telecom-equipment industry, see the second chapter of this report, section headed "India's Electronics Market is as Oligopolized as the Global Industry."

[11] For details, see interview findings in the third chapter.

[12] http://www.slideshare.net/morellimarc/mckinsey-manufacturing-future-2013-22958651 .

[13] Department of Industrial Policy & Promotion, FDI statistics, http://dipp.nic.in/English/Publications/FDI_Statistics/2013/india_FDI_April2013.pdf .

[14] "What Awaits Raghuram Rajan: Falling Rupee, High Current Account Deficit," *Press Trust of India,* September 4, 2013, http://profit.ndtv.com/news/economy/article-what-awaits-raghuram-rajan-falling-rupee-high-current-account-deficit-326617 .

[15] Thomson Reuters Datastream data, quoted in D. Pilling and J. Noble, "Storm Defences Tested—Asia," *Financial Times,* August 29, 2013, p. 5.

[16] *Faster, Sustainable and More Inclusive Growth: An Approach to the 12th Five Year Plan* (New Delhi: Government of India Planning Commission, 2011), p. 80, italics added.

[17] Ibid., p. 80.

[18] P. Marsh, *The New Industrial Revolution: Consumers, Globalization, and the End of Mass Production* (New Haven and London: Yale University Press, 2012). See also G.P. Pisano and W.C. Shih, *Producing Prosperity: Why America Needs a Manufacturing Renaissance* (Boston, MA: Harvard Business Review Press, 2012). For transformations in the electronics industry, see D. Ernst, *A New Geography of Knowledge in the Electronics Industry? Asia's Role in Global Innovation Networks*, Policy Studies, no. 54 (Honolulu: East-West Center, 2009) and D. Ernst, *High Road or Race to the Bottom? Reflections on America's Manufacturing Futures,* East-West Center Working Paper (Honolulu: East-West Center, forthcoming).

[19] S.S. Shipp, N. Gupta, B. Lal, J.A. Scott, C.L. Weber, M.S. Finnin, M. Blake, S. Newsome, S. Thomas, *Emerging Global Trends in Advanced Manufacturing*, report prepared for the Office of the Director of National Intelligence (Washington, DC: Institute for Defense Analyses, 2012), http://www.dtic.mil/dtic/tr/fulltext/u2/a558616.pdf .

[20] For the United States, see *A National Strategic Plan for Advanced Manufacturing* (Washington, DC: National Science and Technology Council, February 2012); *Capturing Domestic Competitive Advantage in Advanced Manufacturing* (Gaithersburg, MD: Advanced Manufacturing Partnership Steering Committee Report to the President, July 2012). See also D. Ernst, *High Road or Race to the Bottom? Reflections on America's Manufacturing Futures,* East-West Center Working Paper (Honolulu: East-West Center, forthcoming).

[21] http://wohlersassociates.com/additive-manufacturing.html .

[22] Richard A. D'Aveni, "3-D Printing Will Change the World" (Boston, MA: *Harvard Business Review,* March 2013), http://hbr.org/2013/03/3-d-printing-will-change-the-world/ar/1 .

[23] E. Anderson, *Additive Manufacturing in China: Threats, Opportunities, and Developments (Part 1)*, SITC Bulletin Analysis, May 2013, quoting "2012 年增材制造技术国际论坛暨第六届全国增材制造技术学术会议在武汉召开" [2012 Additive Manufacturing Technology International Forum and Sixth National Additive Manufacturing Conference Opens in Wuhan], December 28, 2012, http://www.nmgjxw.gov.cn/cms/zbhygzdt/20121228/7986.html .

[24] The United States continues to lead with 38 percent of all industrial 3DP installations, followed by Japan (9.7 percent), Germany (9.4 percent), and China (8.7 percent). "3D Printing Market in China to Reach $ 1.6 Billion Within Three Years," http://www.3ders.org/articles/20130530-3d-printing-market-in-china-to-reach-billion-within-three-years.html .

[25] G. Tromans, "View From the East," *TCT Magazine,* February 21, 2013, http://www.tctmagazine.com/additive-manufacturing/view-from-the-east/ .

[26] E. Anderson, 2013, *Additive Manufacturing in China: Aviation and Aerospace Applications (Part 2)*, http://www-igcc.ucsd.edu/assets/001/504640.pdf .

[27] *Capital Goods: 3D Printing—Don't Believe (ALL) The Hype* (Morgan Stanley Blue Paper, September 5, 2013) and "3D Printing is Here—But the Factory in Every Home isn't Here Yet!" (Deloitte Global Services Ltd., 2012), https://www.deloitte.com/assets/Dcom-Global/Local%20Content/Articles/TMT/TMT%20Predictions%202012/16470A%20lb%203D%20printing%20lb1.pdf. See also B. Garrett, *Could 3D Printing Change the World?* (Washington, DC: Atlantic Council Strategic Foresight Initiative Report, October 17, 2011), http://www.atlanticcouncil.org/blogs/new-atlanticist/will-3d-printing-change-the-world.

[28] *The Manufacturing Plan: Strategies for Accelerating Growth of Manufacturing in India in the 12th Five Year Plan and Beyond* (New Delhi: Government of India Planning Commission, 2012), p. 122, http://planningcommission.gov.in/aboutus/committee/strgrp12/str_manu0304.pdf .

[29] *Faster, Sustainable and More Inclusive Growth: An Approach to the 12th Five Year Plan* (New Delhi: Government of India Planning Commission, 2011), p. 5.

[30] http://www.rdmag.com/articles/2012/12/bric-india .

[31] *The Global Innovation Index 2013: The Local Dynamics of Innovation* (Geneva, Ithaca, and Fontainebleau: Cornell University, INSEAD, and WIPO, 2013).

[32] *India Sustaining High and Inclusive Growth* (Organisation for Economic Co-operation and Development, Paris, 2012), p. 14, http://www.oecd.org/india/IndiaBrochure2012.pdf .

[33] *R&D Magazine* (Battelle Global R&D Funding Forecast), 2013, http://www.rdmag.com/digital-editions/2012/12/2013-r-d-magazine-global-funding-forecast.

[34] James Heitzman and Robert L. Worden, eds., "India: A Country Study," *Science and Technology* (Washington, DC: GPO for the Library of Congress), 1995, http://countrystudies.us/india/101.htm .

[35] G. Balachandran, ed., *India and the World Economy, 1850–1950* (Oxford and New York: Oxford University Press, 2005).

[36] In 1700 India had a 22.6 percent share of the world's GDP, China had 23.1 percent, and the whole of Europe had 23.3. percent (Angus Maddison, *The World Economy: A Millennial Perspective* (Paris: Organisation for Economic Co-operation and Development, 2001).

[37] P. Agarwal, "Higher Education in India: The Need for Change" (working paper 180, Indian Council for Research on International Economic Relations, New Delhi, 2006), p. 40, quoting data from the Government of India Planning Commission's 2002 *Report of Special Group on Targeting Ten Million Employment Opportunities a Year in the Tenth Five-Year Plan.*

[38] B. Jalan, *Future of India: Politics, Economics and Governance* (New Delhi: Penguin, 2005).

[39] Vijay Kelkar, "India: On the Growth Turnpike" (Narayanan Oration, Australian National University, April 2004). See also Vijay Kelkar, "India's Emerging Economic Challenges," *Economic and Political Weekly,* Mumbai, August 14, 1999, pp. 2326–29.

[40] P. Agarwal, "Higher Education in India: The Need for Change" (working paper 180, Indian Council for Research on International Economic Relations, New Delhi, 2006), http://www.icrier.org/pdf/ICRIER _WP180__Higher_Education_in_India_.pdf .

[41] http://www.rdmag.com/articles/2012/12/bric-india .

[42] D. Kapur and B.P. Mehta, *Mortgage the Future? India's Higher Education System* (Washington, DC: Brookings-NCAER India Policy Forum, 2007).

[43] C. Dahlman and A. Utz, *India and the Knowledge Economy: Leveraging Strengths and Opportunities* (Washington, DC: The World Bank, 2005), ch. 3.

[44] Partha Mukhopadhyay, email to author, August 19, 2013.

[45] *Report of the Working Group on Information Technology Sector, Twelfth Five Year Plan* (New Delhi: Ministry of Communications and Information Technology Department of Information Technology, 2012), ch. 5, "e-Industry (Electronics Hardware)," http://planningcommission.gov.in/aboutus/committee/wrkgrp12 /cit/wgrep_dit.pdf .

[46] Report of the Working Group on Information Technology Sector, Twelfth Five Year Plan (New Delhi: Ministry of Communications and Information Technology Department of Information Technology, 2012), ch. 5, "e-Industry (Electronics Hardware)," quoting unpublished background report by P. Sonderegger and N. Mehta, *Tamil Nadu Manufacturing Skills Delivery.*

[47] N. Katoch, L.S. Jordan, and D. Bhattasali, "Regulations Restricting the Growth of Non-Farm Enterprises" (unpublished manuscript, April 3, 2013, New Delhi: The World Bank).

[48] For details on the interview sample, methodology, and findings, see the third and fourth chapters of this study.

[49] http://faculty.chicagobooth.edu/raghuram.rajan/research/papers/A%20hundred%20small%20steps.pdf .

[50] P. Aghion, R. Burgess, S. Redding, and F. Zilibotti, "The Unequal Effects of Liberalization: Evidence from Dismantling the License Raj in India" (working paper 12031, Washington, DC: National Bureau of Economic Research, February 2006), p. 24. Focusing on differences in labor regulations across India's states, these authors argue that "local institutions and policies matter for whether a region benefits or is harmed by nationwide liberalization reforms. This is in line with a small but growing trade liberalization literature which points to heterogeneous effects depending on the local institutional setting in which liberalization takes place."

[51] D. Acemoglu, P. Aghion, and F. Zilibotti, "Distance to Frontier, Selection, and Economic Growth," *Journal of the European Economic Association* 4, no. 1 (March 2006), pp. 37–74.

[52] R. Rajan, "Why India Slowed," April 30, 2013, http://www.project-syndicate.org/print/the-democratic -roots-of-india-s-economic-slowdown-by-raghuram-rajan .

[53] http://business.gov.in/legal_aspects/industries_act.php .

[54] For details see D. Ernst and D. O'Connor, *Competing in the Electronics Industry: The Experience of Newly Industrialising Economies* (Paris: Organisation for Economic Co-operation and Development, Development Centre Studies, 1992); and J. Grieco, *Between Dependency and Autonomy: India's Experience with the International Computer Industry* (Berkeley and Los Angeles: University of California Press, 1984).

[55] P. Evans, *Embedded Autonomy: States and Industrial Transformation* (Princeton, NJ: Princeton University Press, 1995), p. 115.

[56] *Statement on Industrial Policy* (New Delhi: Government of India, Ministry of Industry, July 24, 1991), http://dipp.nic.in/English/Policies/Industrial_policy_statement.pdf .

[57] R. Rajan, *Why India Slowed* (April 30, 2013), http://www.project-syndicate.org/print/the-democratic-roots-of-india-s-economic-slowdown-by-raghuram-rajan .

[58] For research on industrial and innovation policies in Japan, Korea, and Taiwan and their use of policy tools which today would be prohibited by WTO rules, see G. Odagiri, H. Odagiri, and A. Goto, *Technology and Industrial Development in Japan: Building Capabilities by Learning, Innovation, and Public Policy* (Oxford: Oxford University Press, 1996); Linsu Kim, *Imitation to Innovation: The Dynamics of Korea's Technological Learning* (Cambridge, MA: Harvard Business Press, 1997); and Tain-Jy Chen and J.L. Lee, eds., *The New Knowledge Economy of Taiwan* (Cheltenham, UK: Edgar Elgar Publishers, 2004).

[59] For details, http://www.wto.org/english/tratop_e/trims_e.htm .

[60] TRIPS agreement coverage is pervasive and includes copyright and related rights (i.e., the rights of performers, producers of sound recordings, and of broadcasting organizations), trademarks including service marks, geographical indications including appellations of origin, industrial designs, patents including the protection of new varieties of plants, the layout-designs of integrated circuits, and undisclosed information including trade secrets and test data.

[61] TRIPS also provides for certain basic principles such as national and most-favored-nation treatment and general rules ensuring that procedural difficulties in acquiring or maintaining IPRs do not nullify any substantive benefits from the agreement.

[62] World Trade Organization, *A Handbook on the WTO TRIPS Agreement* (Cambridge, UK: World Trade Organization and Cambridge University Press, 2012), http://www.wto.org/english/res_e/publications_e/handbook_wtotripsag12_e.htm .

[63] C. Henry and J. Stiglitz, "Intellectual Property: Dissemination of Innovation and Sustainable Development," *Journal of Global Policy* 1 (2010), pp. 237–51.

[64] J.M. Curtis "Trade and Innovation: Challenges and Policy Options," background paper, ICTSD Expert Group Meeting, Geneva, June 6–7, 2013.

[65] According to *Wikipedia*, the Anti-Counterfeiting Trade Agreement (ACTA) "is a multinational treaty for the purpose of establishing international standards for intellectual property rights enforcement. The agreement aims to establish an international legal framework for targeting counterfeit goods, generic medicines and copyright infringement on the Internet, and would create a new governing body outside existing forums, such as the World Trade Organization, the World Intellectual Property Organization, or the United Nations." The agreement was signed in October 2011 by Australia, Canada, Japan, Morocco, New Zealand, Singapore, South Korea, and the United States. In 2012 Mexico, the European Union, and twenty-two member states of the European Union also signed the agreement.

[66] http://www.wto.org/english/tratop_e/gproc_e/gp_gpa_e.htm .

[67] Construction-services procurements in excess of US$7,777,000 are also subject to the WTO/GPA. This threshold changes every two years.

[68] http://www.ustr.gov/trade-topics/government-procurement/wto-government-procurement-agreement .

[69] *The World Trade Organization/Government Procurement Agreement* (Boston, MA: The Office of the Governor of Massachusetts Frequently Asked Questions, 2012), http://www.mass.gov/anf/docs/osd/pic/wtogpa-faqs-2012.doc .

[70] *The Public Procurement Bill* 2011 (New Delhi: Government of India Planning Commission, 2011), pp. xiii–xiv, http://planningcommission.nic.in/reports/genrep/public_pro_bill.pdf .

[71] Ibid., p. vi. For detailed analysis and policy recommendations, see *Government Procurement in India: Domestic Regulations and Trade Prospects* (New Delhi: CUTS International, published with the support of the British High Commission, 2012).

[72] *The Public Procurement Bill* (New Delhi: Government of India Ministry of Finance Department of Expenditure Procurement Policy Division, 2012), http://www.cci.gov.in/images/media/presentations /MeenaAggarwal10Oct2012.pdf .

[73] *Government Procurement in India: Domestic Regulations and Trade Prospects* (New Delhi: CUTS International, published with the support of the British High Commission, 2012). p. 34.

[74] Soon after ITA became effective in April 1997, participants commenced a schedule of phased duty reductions with all duties slated for elimination by 2000.

[75] For ITA-1 product lists, see Jürgen Richtering, "ITA Products and Harmonized System," WTO Information Technology Symposium, March 29, 2007, p. 12, http://www.wto.org/english/tratop_e /inftec_e/symp_march07_e/richtering_e.ppt#309 .

[76] M. Anderson and J. Mohs, "The Information Technology Agreement: An Assessment of World Trade in Information Technology Products," *International Commerce and Economics, United States International Trade Commission*, 2010, p. 41. http://www.usitc.gov/publications/332/journals/info_tech_agreement.pdf .

[77] India and several other developing countries—including Costa Rica, Indonesia, South Korea, and Taiwan—implemented extended duty staging to 2005 on a product-by-product basis as permitted in the ITA declaration.

[78] According to the Government/Authorities Meeting on Semiconductors, multi-component integrated circuits are a combination of one or more monolithic, hybrid, and/or multi-chip integrated circuits. Products likely to include MCOs are system-on-a-chip, package-on package, and system-in-package devices. See National Research Council, *The New Global Ecosystem in Advanced Computing: Implications for U.S. Competitiveness and National Security* (Washington, DC: The National Academies Press, 2012).

[79] For detailed analysis of the draft expanded product list for ITA-2, see *The Information Technology Agreement: Advice and Information on the Proposed Expansion: Part 1* (Washington, DC: United States International Trade Commission, 2012), http://www.usitc.gov/publications/332/pub4355.pdf and *The Information Technology Agreement: Advice and Information on the Proposed Expansion: Part 2* (Washington, DC: United States International Trade Commission, 2013), http://www.usitc.gov/publications/332/pub4382.pdf .

[80] Implications for India of current negotiations concerning an expansion of ITA are discussed below.

[81] S.J. Ezell, *Boosting Exports, Jobs and Economic Growth by Expanding the ITA* (Washington, DC: The Information Technology and Innovation Foundation, 2012), pp. 8–9.

[82] M. Anderson and J. Mohs, "The Information Technology Agreement: An Assessment of World Trade in Information Technology Products," *International Commerce and Economics, United States International Trade Commission*, 2010, p. 41. http://www.usitc.gov/publications/332/journals/info_tech_agreement.pdf .

[83] Quote from presentation by Greg Slater, Intel's director for trade and competition policy, at "Boosting Exports, Jobs and Economic Growth by Expanding the ITA," Washington, DC, March 15, 2013, hosted by the Information Technology and Innovation Foundation, http://www.itif.org/files/2012-ita-slater.pdf . At the same event, Miriam E. Sapiro, deputy United States trade representative, is quoted as saying "ITA has been one of the most successful agreements ever undertaken in the multilateral trading system" as it has boosted US information technology exports, http://www.itif.org/files/2012-sapiro-comments-ita.pdf .

[84] http://www.aaas.org/spp/rd/presentations/20110218PatrickWilson.pdf . Slide 5; market share based on headquarters of seller.

[85] "Doubling Semiconductor Exports Over the Next Five Years," (San Jose, CA: Semiconductor Industry Association, June 17, 2010), http://www.sia-online.org/clientuploads/directory/DocumentSIA/Export /Doubling_Exports_Paper_0610.pdf .

[86] Ian Steff, Semiconductor Industry Association, testimony in United States International Trade Commission hearing, November 8, 2012, quoted in United States International Trade Commission, *The Information Technology Agreement: Advice and Information on the Proposed Expansion—Part 2,* http://www.usitc.gov/publications/332/pub4382.pdf , pp. 39, 41.

[87] M. Anderson and J. Mohs, "The Information Technology Agreement: An Assessment of World Trade in Information Technology Products," *International Commerce and Economics, United States International Trade Commission,* 2010, p. 17. http://www.usitc.gov/publications/332/journals/info_tech_agreement.pdf .

[88] Ibid., p. 4.

[89] Ibid., p. 7.

[90] http://search.worldbank.org/data?qterm=India%20per%20capite%20GDP%20for%201997&language=EN

[91] M. Anderson and J. Mohs, "The Information Technology Agreement: An Assessment of World Trade in Information Technology Products," *International Commerce and Economics, United States International Trade Commission,* 2010, p. 17. http://www.usitc.gov/publications/332/journals/info_tech_agreement.pdf .

[92] M. Kallummal, "Process of Trade Liberalisation under the Information Technology Agreement: The Indian Experience" (working paper 200/3, Centre for WTO Studies, Indian Institute of Foreign Trade, New Delhi, 2012), p. 15.

[93] The International Harmonized System (HS) trade code 85 covers "electrical machinery and equipment and parts thereof; sound recorders and reproducers, television image and sound recorders and reproducers, and parts," http://www.foreign-trade.com/reference/hscode.cfm?cat=13 .

[94] M. Kallummal, "Process of Trade Liberalisation under the Information Technology Agreement: The Indian Experience" (working paper 200/3, Centre for WTO Studies, Indian Institute of Foreign Trade, New Delhi, 2012), p. 36, fig. 19.

[95] For details, see Directorate General of Foreign Trade website Import Chapter 85-2011 (pdf). http://www.infodriveindia.com/india-trade-data/more-info.aspx

[96] *How Imports Improve Productivity and Competitiveness* (Paris: Organisation for Economic Co-operation and Development, 2010), http://www.oecd.org/trade/45293596.pdf .

[97] M. Kallummal, "Process of Trade Liberalisation under the Information Technology Agreement: The Indian Experience" (working paper 200/3, Centre for WTO Studies, Indian Institute of Foreign Trade, New Delhi, 2012), pp. 17–18.

[98] A. Jauhri, "Implementing TBT Agreement. Indian Experience," presentation at the Indian Institute of Foreign Trade Centre for WTO Studies National Workshop on the WTO Agreement on Technical Barriers to Trade, New Delhi, April 4–5, 2013.

[99] According to the Commerce and Industry Ministry, "India's experience with the ITA-1 has not been encouraging as it has almost wiped out the IT industry from India. After examining the matter in consultation with the nodal Ministry i.e. Department of Electronics and Information Technology and other stakeholders, it has been decided, for the present, not to join the negotiations as it will not be in our national interest." Quoted in "India to Skip Talks on Expanding ITA Scope," *The Hindu,* March 13, 2013.

[100] Multi-component integrated circuits are used in a wide variety of products including smartphones, tablets, medical devices, and household appliances and car parts such as braking, steering, and air-bag systems. Multi-component integrated circuits thus can be classified under a wide range of India's harmonized system of coding subheadings. As a result, no one really knows for sure how important multi-component integrated circuits are for US exports.

[101] According to United States International Trade Commission, major US semiconductor companies producing multi-component integrated circuits in the United States include Intel, Texas Instruments, Freescale, ON Semiconductor, Analog Devices and leading fab-less chip-design companies such as Qualcomm, Broadcomm, Nvidia, and Cypress Semiconductor.

[102] United States International Trade Commission, *The Information Technology Agreement: Advice and Information on the Proposed Expansion—Part 1,* 2013, p. vi.

[103] "Observations on 'Expanding the Information Technology Agreement," based on deliberations at the ASSOCHAM National WTO Council, roundtable held in New Delhi, December 20, 2012, http://view.officeapps.live.com/op/view.aspx?src=http%3A%2F%2Fwww.assocham.org%2Fdocs%2FITA-Roundtable-Observation-Paper_13.2.13.doc .

[104] Ibid., p. 8.

[105] "U.S., Others Suspend ITA Talks to Pressure China to Soften its Stance," *Inside U.S. Trade,* July 29, 2013, p. 2.

[106] For more recent developments, see D. Ernst, "Does the Information Technology Agreement Facilitate Industrial Development and Innovation? India's and China's Diverse Experiences," Think Piece prepared for the E15 Expert Group on Trade and Innovation, co-sponsored by the International Center for Trade and Sustainable Development (Geneva: IMD Lausanne and the Evian Group, forthcoming).

[107] John Neuffer, senior vice president for global policy at the Washington, DC-based Information Technology Industry Council, quoted in "U.S., Others Suspend ITA Talks to Pressure China to Soften its Stance," *Inside U.S. Trade,* July 29, 2013, p. 2.

[108] The WTO defines FTAs as "agreements among two or more parties in which reciprocal preferences (whether or not reaching complete free trade) are exchanged to cover a large spectrum of the parties trade." Customs unions, on the other hand, are PTAs "with a common external tariff in addition to the exchange of trade preferences." Both forms of PTAs can be either bilateral (involving two countries) or plurilateral (involving three or more countries).

[109] See "India's Free Trade Agreements," April 27, 2011, for a list of India's FTAs and PTAs, http://www.india-briefing.com/news/indias-free-trade-agreements-4810.html/ . See, for an analysis of the impact of ITA-1 on India, http://wtocentre.iift.ac.in/workingpaper/Working%20Paper3.pdf .

[110] *India ASEAN Free Trade Agreement: Implications for India's Economy* (Deloitte-FICCI, 2011), p. 61, http://www.deloitte.com/assets/Dcom-India/Local%20Assets/Documents/India_ASEAN_FTA.pdf .

[111] S. Francis, "A Sectoral Impact Analysis of the ASEAN-India Free Trade Agreement," *Economic and Political Weekly* XLVI No.2 (January 8, 2011), p. 52, table 6: India's Tariff Reduction Scenario in Major Sectors Involved in Two-Way Trade with ASEAN.

[112] Ibid., p. 54, table 6: India's Tariff Reduction Scenario in Major Sectors Involved in Two-Way Trade with ASEAN.

[113] P. Draper and M. Dube, *Plurilaterals and the Multilateral Trading System: An Issue Brief for the ICSTD Expert Group on Preferential Trade Agreements* (Geneva: ICSTD and IDB, 2013).

[114] For instance, TPP-2 art. 8.6 of a leaked US TPP Proposal for an IP chapter would require TPP members to grant extensions of patent terms beyond the TRIPS twenty-year minimum patent term to compensate both for delays in patenting and in granting marketing approval. Patent term extensions delay the introduction of generic products into a market, maintaining monopoly protections and higher prices during the extension (S. Flynn, *Public Interest Analysis of the US TPP Proposal for an IP Chapter* (American University Washington College of Law Program on Information Justice and Intellectual Property, December 6, 2011), ch. v: Patent and Data-Related rights. The leaked US TPP Proposal can be accessed at http://keionline.org/sites/default/files/tpp-10feb2011-us-text-ipr-chapter.pdf .

[115] http://en.wikipedia.org/wiki/Trans-Pacific_Partnership_Intellectual_Property_Provisions#cite_note-172 .

[116] In 2012 the top two industry suppliers (Boeing and Airbus) were estimated to account for over 90 percent of all industry revenue. See Hepher, T. and C. Altmeyer, 2013, *"Boeing overtakes Airbus in annual sales race,"* January 17, http://www.reuters.com/article/2013/01/17/us-airbus-orders-idUSBRE90G0CF20130117.

[117] For an economic analysis of the impact of these new chip-design methodologies, see D. Ernst, "Complexity and Internationalization of Innovation: Why is Chip Design Moving to Asia?," *International Journal of Innovation Management* 9 no. 1 (2005a): pp. 47–73.

[118] A.S. Grove, *Only the Paranoid Survive: How to Exploit the Crisis Points that Challenge Every Company and Career* (New York and London: Harper Collins Business, 1996).

[119] The downgrading reflects slower-than-expected sales of Samsung's flagship Galaxy S4 smartphone and declining operating margins for its mobile-devices business as growth in the high-end smartphone market nears saturation.

[120] "Apple Shares Dip on Report it Has Cut iPhone Production," *The Washington Post*, July 9, 2013, http://articles.washingtonpost.com/2013-07-09/business/40460008_1_apple-shares-samsung-and-apple -premium-smartphone-market .

[121] D. Ernst, "Digital Information Systems and Global Flagship Networks: How Mobile is Knowledge in the Global Network Economy?" in *The Industrial Dynamics of the New Digital Economy,* ed. J.F. Christensen (Cheltenham, UK: Edward Elgar, 2003), pp. 151–178.

[122] P. Marsh, "Marvel of the World Brings Both Benefit and Risk," *Financial Times,* June 11, 2010, p. 7. For a detailed case study of the multi-layered global production networks in Asia's electronics industry, see D. Ernst, "Global Production Networks in East Asia's Electronics Industry and Upgrading Perspectives in Malaysia" in *Global Production Networking and Technological Change in East Asia,* eds. Shahid Yusuf, M. Anjum Altaf, and Kaoru Nabeshima (Washington, DC: The World Bank and Oxford University Press, 2004).

[123] D. Ernst, "Innovation Offshoring: Root Causes of Asia's Rise and Policy Implications" in *Multinational Corporations and the Emerging Network Economy in the Pacific Rim*, ed. Juan Palacio (Cheltham, UK: Routledge, 2007).

[124] D. Ernst, *A New Geography of Knowledge in the Electronics Industry? Asia's Role in Global Innovation Networks*, Policy Studies, no. 54 (Honolulu: East-West Center, August 2009).

[125] These locations reflects Huawei's objective to be close to major global centers of excellence and to learn from incumbent industry leaders: Plano, Texas, is one of the leading US telecom clusters centered on Motorola; Kista, Stockholm, plays the same role for Ericsson and, to some degree, Nokia; and the link to British Telecom was Huawei's initial entry into leading global telecom operators.

[126] W.J. Baumol and A.S. Blinder, *Economics: Principles and Policy,* 5th ed. (San Diego: Harcourt Brace Jovanovich, 1991), p. 596.

[127] J. Farrell and C. Shapiro, "Asset Ownership and Market Structure in Oligopoly," *RAND Journal of Economics* 21 (summer 1990), p. 275.

[128] For an economic analysis of innovation capacity, see D. Ernst, *A New Geography of Knowledge in the Electronics Industry? Asia's Role in Global Innovation Networks*, Policy Studies, no. 54 (Honolulu: East-West Center, August 2009).

[129] See the fourth chapter of this study.

[130] D. Ernst and B. Naughton, *Global Technology Sourcing and China's Integrated Circuit Design Industry: A Conceptual Framework and Preliminary Research Findings,* East-West Center Economics Working Paper 131 (Honolulu: East-West Center, August 2012), http://www.EastWestCenter.org/pubs/33626 .

[131] The prices of Chinese quad-core chips for tablets are approximately US$8, which is roughly a third of the price for a comparable chip from a first-tier global industry leader like US-based Nvidia (*Strategy Analytics,* July 2013, http://blogs.strategyanalytics.com/WSS/post/2013/07/26/Strategy-Analytics-Global-Smartphone -Shipments-Hit-Record-230-Million-Units-in-Q2-2013.aspx).

[132] Canalys, July 2013, http://www.canalys.com/newsroom/quarter-billion-smart-phones-ship-q3-2013 .

[133] For an analysis of some of the inherent drawbacks of these policies see D. Ernst, "Is China's Innovation a Serious Challenge?," *Chemistry & Industry* 77 no. 6 (June 2013): p. 44.

[134] J.M. Blair, *Economic Concentration* (New York: Harcourt Brace Jovanovich, 1972).

[135] http://www.economist.com/node/16693547 .

[136] Hewlett-Packard Company, 14.8 percent; Lenovo, 14.7 percent; Dell, 11.0 percent; and Acer, 8.6 percent.

[137] The fifth would be Asus, at 6.8 percent. See Table A-1 in the Appendix.

[138] *IDC Worldwide Quarterly PC Tracker,* July 10, 2013.

[139] In addition to mobile handsets, this includes tablets, ultra notebooks, and laptops.

[140] http://www.economist.com/node/16693547 .

[141] See Appendix, Table A-2 (Worldwide Smartphone Shipments, by OS Vendor, 2012–2017).

[142] See Appendix, Table A-3 (Top Five Smartphone Vendors, Shipments, and Market Share, Q1 2013).

[143] Seagate has 45 percent, Western Digital 42 percent, and Toshiba 13 percent.

[144] *Digital Storage Technology Newsletter* (Coughlin Associates, May 2013). http://www.tomcoughlin.com/. Retrieved July 11, 2013.

[145] For instance, a widely quoted study used the hard-disk drive industry as an example of a highly competitive industry structure (D.G. McKendrick, R.F. Doner, and S. Haggard, *From Silicon Valley to Singapore: Location and Competitive Advantage in the Hard Disk Drive Industry* [Stanford, CA: Stanford University Press, 2000]). For an early analysis of emerging oligopolization trends in the hard-disk drive industry, see D. Ernst, "From Partial to Systemic Globalization: International Production Networks in the Electronics Industry" (working paper 98, Berkeley Roundtable on the International Economy, University of California at Berkeley and the Graduate School of International Relations and Pacific Studies, University of California at San Diego, April 1997, http://brie.berkeley.edu/publications/WP%2098.pdf).

[146] See discussion in the fourth chapter of this study.

[147] http://www.electronics-eetimes.com/en/pace-tops-2009-global-set-top-box-vendor-ranking.html?cmp_id =7&news_id=222902161 .

[148] Internet Protocol set-top boxes have a built-in home-network interface which can either be Ethernet or one of the existing wire home-networking technologies such as HomePNA or the ITU-T G.hn standard, all of which provide a high-speed (up to 1Gbit/s) local area network using existing home wiring (power lines, phone lines, and coaxial cables). "New global standard for fully networked home," ITU-T press release, http://www.itu.int/ITU-T/newslog/New+Global+Standard+For+Fully+Networked+Home.aspx

[149] http://www.infonetics.com/pr/2011/2Q11-STB-Market-Highlights.asp .

[150] http://www.eetasia.com/ART_8800672656_480700_NT_88ff203e.HTM .

[151] International Data Corporation, http://www.crn.in/news/hardware/2013/03/01/indian-pc-market-has -grown-idc .

[152] http://cmrindia.com/more-than-221-million-mobile-handsets-shipped-in-india-during-cy-2012-a-y-o-y -growth-of-20-8-nokia-retains-overall-leadership/ .

[153] Nokia acquired the shares formerly held by Siemens in the Nokia Siemens Networks in July 2013. It is unclear whether Nokia has the financial wherewithal to support its presence in the Indian market.

[154] *Deloitte Report on National Telecom Policy,* quoting from the *Ericsson 2011 Market Report,* http://www.deloitte.com/assets/Dcom-India/Local%20Assets/Documents/Thoughtware/National %20Telecom%20Policy%202011.pdf .

[155] http://businesstoday.intoday.in/story/huawei-zte-market-share-in-india/1/188935.html .

[156] ICRA Limited (formerly Investment Information and Credit Rating Agency of India Limited) Research, http://www.icra.in/AllTypesOfReports.aspx?ReportCategory=Telecom Services .

[157] For a list of the largely low-technology components typically procured by telecom-tower companies from various vendors, see the Indus Towers website http://industowers.com/vendors_questionnaire.php .

[158] On Korea, the classical study remains Linsu Kim, *Imitation to Innovation: The Dynamics of Korea's Technological Learning* (Cambridge, MA: Harvard Business School Press, 1997). On Taiwan, see Tain-Jy Chen and J.L. Lee, eds, *The New Knowledge Economy of Taiwan* (Edgar Elgar Publishers, 2004). On China, see E. Steinfeld, *Playing Our Game* (Oxford University Press, 2010).

[159] S. Berger, "Lessons in Scaling from Abroad: Germany and China," in *Making in America: From Innovation to Market* (Cambridge, MA: The MIT Press, 2013).

[160] As argued in J. Curtis, "Trade and Innovation: Challenges and Policy Options," background paper for the International Centre for Trade and Sustainable Development Expert Group 6 meeting, Geneva, June 6–7, 2013.

[161] http://ictsd.org/publications/latest-pubs/dg2013/mari-pangestu/ .

[162] Growing domestic consumption was highlighted as a key motivation, especially by Indian component producers such as Ace Components, Digital Circuits, ID Smart Cards, SLN Technologies, and Hitech Magnetics.

[163] Note that this contrasts with the current situation in China, where the previously substantial differences in compensation packages between MNCs and large domestic companies such as Lenovo or Huawei has been substantially reduced. Domestic Chinese companies now successfully compete for the best available talent—not just in China but internationally.

[164] This is consistent with the author's research on Intel Bangalore's IC-design projects. Bangalore's contribution to complex multicore processors was mainly in detailed engineering to make those processors "market ready." Tasks performed at Intel Bangalore's laboratory "are mostly tedious and require intense manpower to complete the tasks. The intellectual content, i.e., inventing new techniques and architectures, is low. No new core design was done to complete the project," D. Ernst, *A New Geography of Knowledge in the Electronics Industry? Asia's Role in Global Innovation Networks*, Policy Studies, no. 54 (Honolulu: East-West Center, 2009), p. 21.

[165] D. Bhattasali, N. Katoch, and L.S. Jordan, "Regulations Restricting the Growth of Non-Farm Enterprises," unpublished manuscript, World Bank, New Delhi, April 3, 2013.

[166] See analysis in the second chapter of this study.

[167] *Pre-Budget Recommendations 2013–14* (Electronic Industries Association of India, Bangalore, 2012), http://www.elcina.com/ELCINAs%20Pre%20Budget%20Recommendations%202013-14.pdf .

[168] *Recommendations for Union Budget 2012–13* (Manufacturers' Association for Information Technology, New Delhi, 2012), http://www.capitalmarket.com/Budget/2012-2013/Manufacturers'%20Association%20for%20Information%20Technology%20(MAIT).pdf .

[169] For a detailed analysis of the barriers encountered by Indian medical-device companies, see S. Jaroslawski and G. Saberwal, "Case Studies of Innovative Medical Device Companies from India: Barriers and Enablers to Development," *BMC Health Services Research* 13 (2013): p. 199.

[170] Interviews with Rajoo Goel, secretary general, Electronic Industries Association of India, June 11, 2013, and Syuresh Khanna, secretary general, Consumer Electronics and Appliances Manufacturers Association, June 12, 2013.

[171] Customs Notification 25/99 is available at http://www.cbec.gov.in/customs/cst2012-13/cs-gen/gen-exemptn-idx.htm (point no. 100). More details are available at http://www.cbec.gov.in/customs/cst2012-13/cs-gen/cs-gen99-106.pdf .

[172] For an in-depth analysis of China's standardization strategy, see D. Ernst, *Indigenous Innovation and Globalization: The Challenge for China's Standardization Strategy* (La Jolla, CA: UC Institute on Global Conflict and Cooperation; and Honolulu: East-West Center; 2011; and, in Chinese, Beijing: University of International Business and Economics Press 自主创新与全球化：中国标准化战略所面临的挑战), http://www.EastWestCenter.org/pubs/3904 .

[173] D. Ernst, *America's Voluntary Standards System: A "Best Practice" Model for Asian Innovation Policies*, Policy Studies, no. 66 (Honolulu: East-West Center, 2013), http://www.eastwestcenter.org/pubs/33981.

[174] As defined in *Framework and Roadmap for Smart Grid Interoperability Standards, Release 1.0, Office of the National Coordinator for Smart Grid Interoperability*, National Institute of Standards and Technology Special Publication 1108 (Washington, DC: US Department of Commerce, January 2010).

[175] *A Pilot Study on Technology-Based Start-Ups*, a study prepared for the Government of India Department of Scientific and Industrial Research (New Delhi: Indian Institute of Foreign Trade Centre for International Trade in Technology, 2007), http://www.dsir.gov.in/reports/ittp_citt/Startups.pdf .

[176] Ibid., p. viii.

[177] Ibid.

[178] For Taiwan, see D. Ernst, "Upgrading Through Innovation in a Small Network Economy: Insights from Taiwan's IT Industry," *Economics of Innovation and New Technology* 19 no. 4 (2010): pp. 295–324 and D. Ernst, "Inter-organizational Knowledge Outsourcing: What Permits Small Taiwanese Firms to Compete in the Computer Industry?," *Asia Pacific Journal of Management* 17 no. 2 (August 2000): pp. 223–55. For Korea, see D. Ernst, "Catching-Up and Post-Crisis Industrial Upgrading: Searching for New Sources of Growth in Korea's Electronics Industry," in *Economic Governance and the Challenge of Flexibility in East Asia,* eds. F. Deyo, R. Doner, and E. Hershberg (Lanham, MD, and other locations: Rowman and Littlefield Publishers, 2001) and D. Ernst, *What are the Limits to the Korean Model? The Korean Electronics Industry Under Pressure,* A BRIE Research Monograph, (Berkeley: The Berkeley Roundtable on the International Economy, University of California at Berkeley, 1994).

[179] "India Electronics and Semiconductor Association (IESA) Presents Their Budget Recommendations to the Government," EFY Times.com, February 21, 2013, http://efytimes.com/e1/fullnews.asp?edid=100897.

[180] IMEC International, a world-leading research and development center for nano-electronics, is headquartered in Leuven, Belgium. Its global innovation network includes research and development teams in The Netherlands (Holst Centre in Eindhoven), China, Taiwan, and India (Imec India Private Limited) and offices in Japan and the United States. http://www2.imec.be/be_en/about-imec/company-profile.html.

[181] L.S. Jordan and K. Koinis, "Navigating Asia's Transformation: Techniques for Flexible Policy Implementation," East-West Center Policy Studies, forthcoming.

[182] *The Manufacturing Plan: Strategies for Accelerating Growth of Manufacturing in India in the 12th Five Year Plan and Beyond* (New Delhi: Government of India Planning Commission, 2012), p. 25, http://planningcommission.gov.in/aboutus/committee/strgrp12/str_manu0304.pdf .

[183] Ibid., p. 8.

[184] *World Facts Book* (Washington, DC: Central Intelligence Agency, 2013), https://www.cia.gov/library/publications/the-world-factbook/fields/2119.html . In 2012 Taiwan had slightly more than twenty-three million people while South Korea had about fifty million people.

[185] For an early analysis, see Chi-Ming Hou and San Gee, "National System Supporting Technical Advance in Industry: The Case of Taiwan" in *National Innovation Systems: A Comparative Analysis*, ed. R.R. Nelson (New York and Oxford: Oxford University Press, 1993), ch. 12. For a detailed analysis of Taiwan's industrial-knowledge-network approach, see D. Ernst, "What Permits David to Grow in the Shadow of Goliath? The Taiwanese Model in the Computer Industry" in *International Production Networks in Asia: Rivalry or Riches?,* eds. M. Borrus, D. Ernst, and S. Haggard (London and New York: Routledge, 2000), ch. 5.

[186] http://www.tsia.org.tw/en/tsia_info.php .

[187] http://www.caspa.com/about .

[188] D. Ernst, *Industrial Upgrading through Low-Cost and Fast Innovation: Taiwan's Experience*, East-West Center Economics Working Paper (Honolulu: East-West Center, October 2013).

[189] See http://www.moea.gov.tw/mns/doit_e/content/Content.aspx?menu_id=5442 or http://www.moea.gov.tw/mns/doit_e/content/Content.aspx?menu_id=5438 .

[190] K. Ohno, *Taiwan: Policy Drive for Innovation—Highlights from GRIPS Development Forum Policy Mission* (Tokyo: Japan Graduate Research Institute for Policy Studies, 2011), http://www.grips.ac.jp/vietnam/KOarchives/doc/ES51_ET_taiwan201100517.pdf .

[191] D. Ernst, "Inter-organizational Knowledge Outsourcing: What Permits Small Taiwanese Firms to Compete in the Computer Industry?," *Asia Pacific Journal of Management* 17 no. 2 (August 2000).

[192] http://www.iesaonline.org/downloads/CAREL_Workshop_on_STB_report_final_110612.pdf .

[193] http://www.omeducation.edu.in/admin/images/download/59724_e-Newsletter_DIT_June%202012.pdf .

[194] The latter two clusters are in Andhra Pradesh, rather than Tamil Nadu or Karnataka, due to substantial land-price differences.

[195] http://www.elcina.com/about.asp .

---

[196] http://iesaonline.org/aboutus/index.html .

[197] K. Blind, A. Jungmittag, and A. Mangelsdorf, *The Economic Benefits of Standardization* (Berlin: DIN German Institute for Standardization, 2011). Similar findings are reported for Australia, Canada, France, New Zealand, and the United Kingdom.

[198] On the US standards systems, see D. Ernst, America's Voluntary Standards System–A "Best Practice" Model for Asian Innovation Policies, Policy Studies #66, March 2013, East-West Center, Honolulu, USA, http://www.eastwestcenter.org/pubs/33981. China's standards system is examined in P. Wang, "Global ICT Standards Wars in China, and China's Standard Strategy," manuscript, China National Institute for Standardization, Beijing, 2013; D. Ernst, *Indigenous Innovation and Globalization: The Challenge for China's Standardization Strategy* (La Jolla, CA: UC Institute on Global Conflict and Cooperation; and Honolulu: East-West Center; 2011; and, in Chinese, Beijing: University of International Business and Economics Press 自主创新与全球化：中国标准化战略所面临的挑战), http://www.EastWestCenter.org/pubs/3904; and R.P. Suttmeier, S. Kennedy, and J. Su, *Standards, Stakeholders, and Innovation: China's Evolving Role in the Global Knowledge Economy* (National Bureau of Asian Research, September 2008).

[199] "The Internet of Everything" brings together people, process, data, and things to enhance the relevance and productivity of networked connections—turning information into actions creating new capabilities, richer experiences, and unprecedented economic opportunity for countries, businesses, communities, and individuals.

[200] J. Palfrey and U. Gasser, *Interop: The Promise and Perils of Highly Interconnected Systems* (New York: Basic Books, 2012).

[201] *Framework and Roadmap for Smart Grid Interoperability Standards, Release 1.0, Office of the National Coordinator for Smart Grid Interoperability,* National Institute of Standards and Technology Special Publication 1108 (Washington, DC: US Department of Commerce, January 2010), pp. 19–20.

[202] See D. Ernst, "Complexity and Internationalization of Innovation: Why is Chip Design Moving to Asia?" *International Journal of Innovation Management* 9 no. 1 (2005a): pp. 47–73 and D. Ernst, "Limits to Modularity: Reflections on Recent Developments in Chip Design," *Industry and Innovation* 12 no. 3 (2005b): pp. 303–35.

[203] For details see D. Ernst, *Indigenous Innovation and Globalization: The Challenge for China's Standardization Strategy* (La Jolla, CA: UC Institute on Global Conflict and Cooperation; and Honolulu: East-West Center; 2011, ch. 3, pp. 49 ff.; and, in Chinese, Beijing: University of International Business and Economics Press 自主创新与全球化：中国标准化战略所面临的挑战), http://www.EastWestCenter.org/pubs/3904 .

[204] Interviews with leading standards experts in the United States, the European Union, and China.

[205] Y.A. Pai, "The International Dimension of Proprietary Technical Standards: Through the Lens of Trade, Competition Law and Developing Countries," *Law, Policy & Economics of Technical Standards eJournal* 1, no. 1, March 13, 2013, p. 5.

[206] On Taiwan, see *Taiwan's National Standards—An Overview,* The Bureau of Standards, Metrology and Inspection, http://ita.doc.gov/td/standards/Markets/East%20Asia%20Pacific/Taiwan/Taiwan.pdf ; on Korea: H. Lee and J. Huh, "Korea's Strategies for ICT Standards Internationalisation: A Comparison with China's," *International Journal of IT Standards and Standardization Research* 10 no. 2 (2012); on China, see D. Ernst, *Indigenous Innovation and Globalization: The Challenge for China's Standardization Strategy* (La Jolla, CA: UC Institute on Global Conflict and Cooperation; and Honolulu: East-West Center; 2011; and, in Chinese, Beijing: University of International Business and Economics Press 自主创新与全球化：中国标准化战略所面临的挑战), http://www.EastWestCenter.org/pubs/3904.

[207] T. Ramakrishna, S.K. Murthy, and S. Malhotra, *Intellectual Property Rights and ICT Standards in India*, commissioned by the US National Academies of Science Board of Science, Technology, and Economic Policy (STEP) Project on Intellectual Property Management in Standard Setting Processes, 2012, http://sites.nationalacademies.org/xpedio/idcplg?IdcService=GET_FILE&dDocName=PGA_072484&RevisionSelectionMethod=Latest .

[208] www.bis.org.in .

[209] Per Standard Administration of China (SAC), this compares with more than twenty-two thousand Chinese national standards, www.sac.gov.cn .

[210] http://dst.gov.in/ .

[211] www.qcin.org .

[212] www.nabl-india.org .

[213] ISO/IEC 17025: "General requirements for the competence of testing and calibration laboratories" is the main ISO/CASCO standard used by testing and calibration laboratories.

[214] www.qcin.org/html/nabcb/index.htm .

[215] www.qcin.org/html/nqc/nqc.htm .

[216] http://www.bis.org.in/sf/compltd.pdf .

[217] www.tec.gov.in .

[218] http://www.gisfi.org/ .

[219] http://dosti.org.in/ .

[220] European Committee for Standardization (CEN), the European Committee for Electrotechnical Standardization (CENELEC), and the European Telecommunications Standards Institute (ETSI).

[221] http://www.cencenelec.eu/News/Videos/Pages/vo-2013-001.aspx .

[222] *Report of the Working Group on Information Technology Sector, Twelfth Five Year Plan* (New Delhi: Ministry of Communications and Information Technology Department of Information Technology, 2012), p. 124, http://planningcommission.gov.in/aboutus/committee/wrkgrp12/cit/wgrep_dit.pdf .

[223] INR—USD exchange rate of 0.0153 as of 15:52:20, July 9, 2013; 2.75 Indian Rupee (INR) was equal to 0.0421 United States Dollar (US$).

[224] http://electronicstds.gov.in/CREITG/app_srv/tdc/gl/jsp/readmore.jsp .

[225] *The Manufacturing Plan: Strategies for Accelerating Growth of Manufacturing in India in the 12th Five Year Plan and Beyond* (New Delhi: Government of India Planning Commission, 2012), http://planningcommission.gov.in/aboutus/committee/strgrp12/str_manu0304.pdf .

[226] *Faster, Sustainable and More Inclusive Growth: An Approach to the 12th Five-Year Plan* (New Delhi: Government of India Planning Commission, 2011), p. 84.

[227] Ibid., p. 83.

[228] http://www.trai.gov.in/WriteReadData/userfiles/file/NTP%202012.pdf .

[229] http://www.trai.gov.in/WriteReadData/userfiles/file/NTP%202012.pdf : p. 6.

[230] *Report of the Working Group on Information Technology Sector, Twelfth Five Year Plan* (New Delhi: Ministry of Communications and Information Technology Department of Information Technology, 2012), pp. 110–111, http://planningcommission.gov.in/aboutus/committee/wrkgrp12/cit/wgrep_dit.pdf .

[231] NPE notification quoted in *Directory of Indian Electronics Industry 2013* (New Delhi: Electronic Industries Association of India, 2013), pp. 50–56.

[232] For details, see discussion below in the section headed "The Semiconductor Wafer Fab Policy."

[233] *Report of the Working Group on Information Technology Sector, Twelfth Five Year Plan* (New Delhi: Ministry of Communications and Information Technology Department of Information Technology, 2012), p. 111, http://planningcommission.gov.in/aboutus/committee/wrkgrp12/cit/wgrep_dit.pdf.

[234] *Indian ESDM Market (2011–2015)* (Frost and Sullivan and Indian Electronics and Semiconductor Association (IESA), 2013), p. x.

[235] Ibid., p. xx.

[236] http://www.nti.org/facilities/44/#sthash.sCInMplj.dpuf.

[237] India's Department of Information Technology received three proposals under the 2007 fab policy, including the first-ever LCD-panel unit in India to be set up by Videocon with an investment of US$1.8 bn. The other two were proposals for PV-cell manufacturing units by Moser Baer (US$3.2 bn) and Titan Energy Systems (US$1.2 bn). The department also received fourteen enquiries of which three-to-four were for chip manufacturing. http://articles.economictimes.indiatimes.com/2007-12-26/news/27683916_1_fab-unit-fab -policy-semiconductor-policy.

[238] "India Snoozed, Lost Intel Chip Plant," http://www.forbes.com/2007/09/06/intel-india-china-markets -equity-cx_rd_0906markets1.html.

[239] *Empowered Committee for Identifying Technology and Investors For Setting Up of Semiconductor Wafer Fabrication (Fab) Manufacturing Facilities in the Country* (New Delhi: DEITy, 2011), http://deity.gov.in /sites/upload_files/dit/files/PressRelease.pdf.

[240] Ibid., p. 2.

[241] Ibid., pp. 2–3.

[242] Ibid., p. 3.

[243] Farhang Shadman, director of the University of Arizona's specialized semiconductor research lab, quoted in: http://www.gereports.com/ultrapure-water-for-ultra-advanced-semiconductor-fab/.

[244] "India faces a crippling water crisis," *Deutsche Welle,* May 29, 2013, http://www.dw.de/india-faces-a -crippling-water-crisis/a-16844835.

[245] In a recent discussion on Bangalore's water crisis, the chairman of the Bangalore Water Supply and Sewage Treatment Board, warned: "If you are taking a property in Bengaluru, especially in the peripheral areas, take at your own risk! We really don't have water for those areas." http://globalvoicesonline.org/2013 /07/26/water-shortage-crisis-looms-large-in-bangalore-india/.

[246] I.S.N. Prasad, principal secretary, IT, Biotechnology and Science and Technology, quoted in *The Hindu,* August 22, 2013, http://www.thehindu.com/news/cities/bangalore/bangalore-out-of-electronic-chip -fabrication-unit-race/article5048379.ece.

[247] For an analysis of toxic chemicals used in semiconductor production, see J. Holden and C. Kelty, *The Environmental Effects of the Manufacturing of Semiconductors,* http://cnx.org/content/m14503/latest/ . This report was developed as part of a Rice University class in "Nanotechnology: Content and Context" initially funded by the National Science Foundation under Grant No. EEC-0407237.

[248] "Approved: India to Get Two Chip Fabs," *EET India,* September 13, 2013.

[249] Presentation by Brian Krzanich, then senior vice president and chief operating officer, Intel, at Intel "Investor Meeting 2012." As of this publication, Krzanich is Intel's chief executive officer. http://intelstudios.edgesuite.net/im/2012/pdf/2012_Intel_Investor_Meeting_Krzanich.pdf.

[250] *Modified Special Incentive Package Scheme (M-SIPS)* (New Delhi: DEITy, July 27, 2012), http://deity.gov.in/sites/upload_files/dit/files/MSIPS%20Notification.pdf.

[251] *Electronics Manufacturing Cluster Scheme (EMC)* (New Delhi: DEITy, 2012), http://deity.gov.in/sites /upload_files/dit/files/Scan_EMC-Notification-Gazette.pdf ; *Guidelines for Operation of Electronics Manufacturing Cluster Scheme (EMC)* (New Delhi: DEITy, 2013), http://deity.gov.in/sites/upload_files /dit/files/EMC-Guidlines_Final.pdf.

[252] For regulations on brownfield clusters and for a list of probable greenfield clusters, see *Guidelines for Notifying Brownfield Clusters under M-SIPS* (New Delhi: DEITy, 2013), http://deity.gov.in/sites/upload _files/dit/files/Guidelines%20for%20notifying%20Brownfield%20Clusters%20under%20M-SIPS,%20deity %20,2013.pdf ; and *List of Probable Greenfield Clusters,* (New Delhi: DEITy, 2013), http://deity.gov.in /sites/upload_files/dit/files/Probable%20Greenfield%20clusters(2).pdf.

[253] *Electronics Manufacturing Cluster Scheme (EMC),* (New Delhi: DEITy, 2012), http://deity.gov.in/sites /upload_files/dit/files/Scan_EMC-Notification-Gazette.pdf.

[254] L.S. Jordan and Y. Saleman, *The Implementation of Industrial Parks: Some Lessons Learned in India* (Washington, DC: The World Bank, forthcoming).

[255] "Draft Project Report on the 'Electronics Development Fund' (EDF)" (New Delhi: DEITy, November 20, 2012), http://www.ipcaindia.org/pdffiles/draft%20DPR_version%20Nov2_11192011_AK.pdf.

[256] Ibid., pp. 9 ff.

[257] D. Ernst, *Is the Information Technology Agreement (ITA) Facilitating Latecomer Manufacturing and Innovation? India's Experience,* East-West Center Economics Working Paper 135 (Honolulu: East-West Center, November 2013).

[258] According to Mari Pangestu, a prominent trade economist from Indonesia, plurilateral agreements "should promote economic and technical cooperation recognising the different stages of development of participants. Special and differential treatment can be justified in circumstances where participants face challenges in benefitting from an increase in trade," http://ictsd.org/publications/latest-pubs/dg2013/mari-pangestu/.

[259] http://deity.gov.in/sites/upload_files/dit/files/Foreign%20Trade%20Policy%20(2012)%20(w_e_f_%2005_06_2012)%20(651%20MB).pdf.

[260] Ibid., p. 14.

[261] Market Development Assistance scheme of India's Department of Commerce, http://commerce.nic.in/trade/international_tpp_cis_5.asp.

Appendix

Table A-1. Preliminary Worldwide PC-vendor Unit-shipment Estimates for Q1 2013

Company	Q1 2013 Shipments	Q1 2013 Market Share (%)	Q1 2012 Shipments	Q1 2012 Market Share (%)	Q1 2012–Q1 20123 Growth (%)
HP	11,687,778	14.8	15,301,906	17.2	-23.6
Lenovo	11,666,400	14.7	11,652,664	13.1	0.1
Dell	8,734,892	11.0	9,838,121	11.0	-11.2
Acer Group	6,843,184	8.6	9,582,046	10.9	-29.3
Asus	5,360,470	6.8	5,552,329	6.2	-3.5
Others	34,914,286	44.1	37,170,712	41.6	-6.1
Total	79,207,010	100.0	89,097,778	100.0	-11.1

Note: Data includes desk-based PCs and mobile PCs including mini-notebooks but not media tablets such as the iPad.

Source: "Gartner Says Worldwide PC Shipments in the First Quarter of 2013 Drop to Lowest Levels Since Second Quarter of 2009" Gartner (April 10, 2013) http://www.gartner.com/newsroom/id/2420816. Used with permission.

Table A-2. Worldwide Smart Phone Shipments

OS Vendor	2012	2017
Android	67.7%	67.1%
Apple	19.5%	14.1%
Microsoft	2.4%	12.7%
BlackBerry	4.8%	4.6%
Others	5.6%	1.5%
Grand Total	100%	100%

Source: http://www.canalys.com/newsroom/over-1-billion-android-based-smart-phones-ship-2017. Used with permission.

Table A-3. Top Five Smartphone Vendors, Shipments, and Market Share, Q1 2013 (in millions of units)

Vendor	Q1 2013 Unit Shipments	Q1 2013 Market Share	Q1 2012 Unit Shipments	Q1 2012 Market Share	Year-over-year Change
Samsung	70.7	32.7%	44.0	28.8%	60.7%
Apple	37.4	17.3%	35.1	23.0%	6.6%
LG	10.3	4.8%	4.9	3.2%	110.2%
Huawei	9.9	4.6%	5.1	3.3%	94.1%
ZTE	9.1	4.2%	6.1	4.0%	49.2%
Others	78.8	36.4%	57.5	37.7%	37.0%
Total	216.2	100.0%	152.7	100.0%	41.6%

Note: Data are preliminary and subject to change. Vendor shipments are branded shipments and exclude OEM sales for all vendors.

Source: IDC Worldwide Mobile Phone Tracker, April 25, 2013; http://www.idc.com/getdoc.jsp?containerId=prUS24085413. Used with permission.

Acknowledgments

This study is based on a report commissioned by the World Bank on behalf of the chief economic advisor, Government of India, Raghuram Rajan (now the governor of the Reserve Bank of India). At the World Bank, I am grateful for support from Vincent Palmade, lead economist for South Asia; Ganesh Rasagam, senior private sector development specialist, Africa; and Bertine Kamphuis, private sector development specialist for the World Bank in New Delhi, India. I owe a special debt of gratitude to Luke S. Jordan, formerly a private sector development specialist for the World Bank in New Delhi, India, for his guidance and for sharing his deep knowledge of the dynamics of policymaking in India. At the East-West Center, I am grateful to Charles E. Morrison, president; Nancy D. Lewis, director of research; and Ralph R. Carvalho, special assistant to the president; for supporting this research, and to Elisa W. Johnston, Carol Wong, and Sharon Shimabukuro for fast and effective publication of this study. In India, Sanket Gupta and Mrigank Gutgutia at RedSeer Consulting provided excellent research and logistical support.

Comments are gratefully acknowledged from Ahmed Abdel Latif (International Centre for Trade and Sustainable Development), Suzanne Berger (Massachusetts Institute of Technology, Production in the Innovation Economy Program), Carlos A. Primo Braga (The Evian Group at the Institute for Management Development, Lausanne), Jorge Contreras (American University Washington College of Law), Mark Dutz (World Bank Economic Policy and Debt Department, Poverty Reduction and Economic Management Network), Padmashree Gehl Sampath (Science and Technology Section, United Nations Conference on Trade and Development), Rana Hasan (Asian Development Bank, New Delhi), Derek L. Hill (National Science Foundation), Partha Mukhopadhyay (Centre for Policy Research, New Delhi), Marcus Noland (Peterson Institute for International Economics and East-West Center), Peter Petri (Brandeis University and East-West Center), Michael G. Plummer (Johns Hopkins University School of Advanced International Studies, Bologna, and East-West Center), Pedro Roffe (International Centre for Trade and Sustainable Development), Willy Shih (Harvard Business School), and Mark Wu (Harvard Law School).

About the Author

Dieter Ernst, an East-West Center senior fellow, is an authority on global production networks and the internationalization of research and development in high-tech industries, with a focus on standards and intellectual property rights. His research examines corporate innovation strategies and innovation policies in the United States and in China, India, and other emerging economies. The author has served as a member of the United States National Academies "Committee on Global Approaches to Advanced Computing"; senior advisor to the Organisation for Economic Co-operation and Development, Paris; research director of the Berkeley Roundtable on the International Economy at the University of California at Berkeley; professor of international business at the Copenhagen Business School; and scientific advisor to governments, private companies, and international institutions.

His publications include *Fast-Tracking India's Electronics Manufacturing Industry: Business Environment and Industrial Policy,* a report prepared for the World Bank (forthcoming); *Is the Information Technology Agreement (ITA) Facilitating Latecomer Manufacturing and Innovation? India's Experience* (2013); *Industrial Upgrading through Low-Cost and Fast Innovation—Taiwan's Experience* (2013); *Standards, Innovation, and Latecomer Economic Development—A Conceptual Framework* (2013); *America's Voluntary Standards System: A "Best Practice" Model for Asian Innovation Policies* (2013); *Indigenous Innovation and Globalization: The Challenge for China's Standardization Strategy* (2011; now also published in Chinese); *China's Innovation Policy Is a Wake-Up Call for America* (2011); *A New Geography of Knowledge in the Electronics Industry? Asia's Role in Global Innovation Networks* (2009): *Can Chinese IT Firms Develop Innovative Capabilities within Global Knowledge Networks?* (2008); *China's Emerging Industrial Economy: Insights from the IT Industry* (with Barry Naughton, 2007); *Innovation Offshoring: Asia's Emerging Role in Global Innovation Networks* (2006); "Complexity and Internationalization of Innovation: Why is Chip Design Moving to Asia?" *International Journal of Innovation Management* (2005); *International Production Networks in Asia: Rivalry or Riches?* (2000); and *Technological Capabilities and Export Success: Lessons from East Asia* (1998).

www.ingramcontent.com/pod-product-compliance
Lightning Source LLC
Chambersburg PA
CBHW051338200326

41519CB00026B/7471